COMMUNITY ATTITUDES
TOWARD POLLUTION

by

Susan L. Caris

Cook College, Rutgers, The State University of New Jersey

THE UNIVERSITY OF CHICAGO
DEPARTMENT OF GEOGRAPHY
RESEARCH PAPER 188

1978

Copyright 1978 by Susan L. Caris
Published 1978 by The Department of Geography
The University of Chicago, Chicago, Illinois

Library of Congress Cataloging in Publication Data

Caris, Susan L., 1950–
    Community attitudes toward pollution.
    (Research Paper—The University of Chicago, Department of Geography; no. 188)
    Bibliography: p. 199

    1. Pollution—Economic aspects—Illinois—Chicago—Public opinion. 2. Consumer protection—Illinois—Chicago—Public opinion. 3. Public opinion—Illinois—Chicago. I. Title. II. Series: Chicago. University. Dept. of Geography. Research paper; no. 188.
H31.C514 no. 188 [HC108.C4] 910s [301.15'43'628168097731'1]
ISBN 0-89065-095-0
78-13164

Research Papers are available from:
The University of Chicago
Department of Geography
5828 S. University Avenue
Chicago, Illinois 60637
Price: $6.00 list; $5.00 series subscription

To M.E.

TABLE OF CONTENTS

LIST OF TABLES viii

LIST OF FIGURES x

ACKNOWLEDGMENTS xi

PART ONE

Chapter
I. INTRODUCTION AND RESEARCH DESIGN 3

II. REVIEW OF THE LITERATURE 6

      Discussion of Sample Studies
        Air Pollution
        Water Pollution
        Noise Pollution
        Solid Waste, Pesticides, and Hazardous Chemicals
        Multiple-Pollutant Studies
      Summary of Research to Date
      Problems and Limitations

III. ATTITUDE THEORY 20

      Basic Terms Defined
        Attitude
        Beliefs
        Behavioral Intentions
        Behavior
      Theories of Attitude Acquisition and Change
        Learning Theories
        Heider's Balance Theory
        Osgood and Tannenbaum's Congruity Theory
        Rosenberg's Affective-Cognitive Consistency Theory
        Festinger's Dissonance Theory
        Expectancy Value Theory
        Fishbein's Model for the Prediction of Intention
      Theory Evaluation and Rationale
      Content of Previous Attitude Studies in Light
        of Attitude Theory

IV. APPLICATION OF THEORY: SAMPLE DESIGN, QUESTIONNAIRE
    CONSTRUCTION, AND ADMINISTRATION OF THE QUESTIONNAIRE . 51

      Questionnaire Construction
      From Theory to Practice
        Attitudes toward Pollution
        Attitudes toward Pollution Abatement
        Influence of the Social Environment on
          Behavioral Intent or Social Norms
        Intention to Perform the Behavior in Question
        Actual Behavior
      Sample Design
      Questionnaire Administration

PART TWO

Chapter	Page
V. FREQUENCY AND CONSISTENCY OF QUESTIONNAIRE RESPONSES	69

    Frequency of Responses
      Worry about Various Social Problems
      Seriousness of the Pollution Problem
      Willingness to Abate Pollution
      Actual Behavior with Regard to Pollution Abatement
      Reasons for Not Doing More
      Willingness to Abate Solid Waste Pollution
      Actual Behavior with Regard to Abating Solid Waste Pollution
      Reasons for Not Doing More to Abate Solid Waste Pollution
      Other Reasons for Not Doing More about Solid Waste Pollution
      Willingness to Respond in More Detail
    Consistency between Worry Rate, Perceived Seriousness, and Intent to Abate Pollution
      Worry Rate and Perceived Seriousness of Pollution
      Perceived Seriousness of Pollution and Intent to Abate Pollution
      Worry Rate and Intent to Abate Pollution
      Perceived Seriousness of Pollution and Specific Actions to Abate Pollution

| VI. ATTITUDES, SOCIAL DIFFERENTIATION, AND ENVIRONMENTAL QUALITY . . . . | 104 |

    Attitudes toward Pollution and Social Differentiation
    Attitudes toward Pollution Abatement and Social Differentiation
    Attitudes toward Pollution and Environmental Quality
    Attitudes toward Pollution Abatement and Environmental Quality
    Attitudes toward Specific Pollutants and Environmental Quality
      Attitudes toward Air Pollution and Air Quality
      Attitudes toward Water Pollution and Water Quality
      Attitudes toward Noise Pollution and Noise Levels
      Attitudes toward Solid Waste Pollution and Solid Waste Levels
    Attitudes toward Specific Pollution Abatement Strategies and Environmental Quality
      Attitudes toward Air Pollution Abatement and Air Quality
      Attitudes toward Water Pollution Abatement and Water Quality
      Attitudes toward Noise Pollution Abatement and Noise Levels
      Attitudes toward Solid Waste Pollution Abatement and Solid Waste Levels

| VII. PREDICTION OF BEHAVIORAL INTENT AND BEHAVIOR . . . . . | 131 |

    Prediction of Behavioral Intent to Abate Pollution
    Prediction of Behavioral Intent to Abate Solid Waste Pollution

Chapter	Page
VIII. CONCLUSIONS AND OPPORTUNITIES FOR FUTURE RESEARCH	146

APPENDIXES 151

   1. LITERATURE REVIEW 153

   2. QUESTIONNAIRE 162

   3. THE CHICAGO STUDY AREAS 166

   4. THE MEASURES OF ENVIRONMENTAL QUALITY 182

SELECTED BIBLIOGRAPHY 199

LIST OF TABLES

1. Interview Response Rates by Community 63
2. Number of Interviews Needed, by Community 65
3. Citywide Frequency of Worry about Various Social Problems 70
4. Seriousness of the Pollution Problem 72
5. Seriousness of Specific Pollution Problems 73
6. Ranking of Specific Actions to Abate or Avoid Pollution 74
7. Ranking of Specific Actions to Abate Solid Waste Pollution 77
8. Worry and Perceived Seriousness of Pollution 89
9. Summary Statistics: Frequency of Worry and Perceived Seriousness of Pollution 90
10. Worry and Perceived Seriousness of Air Pollution 91
11. Summary Statistics: Frequency of Worry and Perceived Seriousness of Specific Pollution 92
12. Worry and Perceived Seriousness of Water Pollution . . . 93
13. Worry and Perceived Seriousness of Noise Pollution . . . 94
14. Worry and Perceived Seriousness of Solid Waste Pollution 95
15. Perceived Seriousness of Pollution and Intent to Abate Pollution 96
16. Summary Statistics: Perceived Seriousness of Pollution and Intent to Abate Pollution 97
17. Worry about Pollution and Intent to Abate Pollution 98
18. Summary Statistics: Rate of Worry and Intent to Abate Pollution 99
19. Ranking of Specific Actions to Abate or Avoid Pollution by Perceived Seriousness of the Problem 100
20. Ranking of Specific Actions to Abate or Avoid Pollution, by Control Categories 102

21.	Mean Attitudes toward Pollution and Pollution Abatement	106
22.	Social Variables	108
23.	Spearman Rank-Order Correlation Matrix: Attitude toward Pollution and Social Characteristics	111
24.	Spearman Rank-Order Correlation Matrix: Attitude toward Pollution Abatement and Social Characteristics	114
25.	Environmental Quality Variables	116
26.	Spearman Rank-Order Correlation Matrix: Attitude toward Pollution and Environmental Quality	120
27.	Spearman Rank-Order Correlation: Attitudes toward Pollution Abatement and Environmental Quality	123
28.	Spearman Rank-Order Correlation: Attitudes toward Solid Waste Pollution and Levels of Solid Waste Pollution	126
29.	Pearson Product-Moment Correlation Coefficients	134
30.	Least-Squares Regression of Intent to Abate Pollution on Attitudes toward Abatement and Subjective Norms	136
31.	Least-Squares Regression of Intent to Abate Pollution on Attitudes toward Pollution Abatement and Subjective Norms	138
32.	Pearson Product-Moment Correlation Coefficients: Solid Waste	140
33.	Least-Squares Regression of Intent to Abate Solid Waste Pollution on Attitudes toward Solid Waste Abatement and Subjective Norms (By Type)	142
34.	Least-Squares Regression of Intent to Abate Solid Waste Pollution on Attitudes toward Solid Waste Abatement and Subjective Norms (By Area)	144

LIST OF FIGURES

1. Balanced and imbalanced triads 29
2. Community areas in Chicago 167
3. Particulates, 1974 184
4. Particulates, 1975 184
5. Cumulative air pollution index 185
6. Air pollution index for 1975 187
7. Index of trace metals in the atmosphere 188
8. Water pollution index 191
9. Noise pollution index 193
10. Solid Waste pollution index 195

ACKNOWLEDGMENTS

Many people were instrumental in focusing my attention on this topic. First is Brian J. L. Berry, who, through his research on the social burdens of pollution, stimulated my latent interest in the behavioral implications of this topic. Stephen Golant guided my reading in this area and provided valuable advice. Donald Jones served as a sounding board, contributing ideas and criticisms. Any errors and omissions, however, are my own.

Many other people at the University of Chicago deserve mention. Kathleen Zar, geography bibliographic librarian, was of great help in my search through the literature, always finding the seemingly impossible reference and source. My fellow graduate students provided a stimulating atmosphere in which to complete this work, and I owe the following individuals thanks for making the long summer nights bearable and productive: Thomas Tocalis, Gundars Rudzitis, Christopher Müller-Wille, Christopher Saricks, and Sherry Sidhu.

Financial assistance for the attitudinal survey was provided by the University of Chicago's Center for Urban Studies and the Department of Geography, and by E. M. Cutter. Richard Jaffee of the Institute for Social Action deserves warm thanks for his help and guidance in the administration of the survey and the reinforcement he gave me regarding the survey design.

I also wish to thank my colleagues at the University of Washington for their encouragement and for the pleasant atmosphere in which to finish the final draft of this manuscript. I would most like to thank the graduate students at the University of Washington for making my stay a very enjoyable experience and for keeping me on my toes intellectually.

   Most of all I wish to thank the residents of Chicago who so enthusiastically responded to this survey on their attitudes toward pollution and pollution abatement.

Seattle
June 1977

PART ONE

CHAPTER I

INTRODUCTION AND RESEARCH DESIGN

Since Earth Day 1970 there has been an increasing interest in environmental quality, not only among the public but also on the part of government and academicians. A wealth of information has been written about the various aspects of environmental quality in urban contexts, ranging from the social burdens of environmental pollution (Berry 1977) to the quality of life, which includes social as well as environmental indicators (Bauer 1966; Liu 1975).

With the advent of federal legislation on air and water pollution, local, state, and federal authorities have shown new interest in planning to improve environmental quality. One of the main questions that consistently appears is, What are the attitudes toward environmental pollution of the people we are trying to protect? Are they willing to accept the programs we design and implement? Do these attitudes vary by sector of the country (rural vs. urban), and are there differences between urban areas?

The main aim of the present research is to examine community attitudes toward pollution and to determine what coping strategies or actions community residents choose to deal with the problem of environmental pollution. Variations in attitudes toward pollution and attitudes toward pollution abatement among social groups are examined, as well as different levels of pollution. The study of these attitudes toward pollution is then placed in a broader framework involving decision-making and the question whether federal and state environmental quality standards should be responsive to community attitudes and community willingness for action.

A rationale for a study of this type is provided by Shusky et al. (1964), who contend that the study of attitudes in relation to comprehensive community environmental pollution analysis is needed to determine the character and significance of the problem in the community as seen by the residents; provide guidance in the development of public information and education programs; and determine the extend to which the public is prepared to accept and support various elements of a control program. The last point is particularly important for this study. The public should contribute to the design of such programs, and we need to know their concerns. As White (1966, p. 109) points out: "Strictly speaking, there is no single expert opinion about attitudes toward quality of environment; there are the opinions each person holds, the opinions he thinks others hold, and the opinions he thinks they should hold. Many public administrators get mixed up about this."

Chapter 2 of this book discusses the relevant literature. Chapter 3 focuses on attitude theory in general, the various types of theories developed by psychologists, and the particular attitude theory chosen as the theoretical foundation for this study. Chapter 4 discusses the process of designing the questionnaire, application of the attitude theory, and the implementation of the questionnaire. Part Two discusses the results of the study: chapter 5 presents the results of the questionnaire, chapter 6 discusses attitudes, social differentiation, and environmental quality, and chapter 7 covers prediction of intent to abate pollution. Finally, chapter 8 summarizes the conclusions drawn from the study and suggests some future topics for more detailed research.

CHAPTER II

REVIEW OF THE LITERATURE

Many disciplines have contributed both to general knowledge about pollution and to theories about how people perceive pollution. Under the general heading of environmental perception, psychology, sociology, and geography have made the most significant contributions (Craik 1973; Proshansky, Ittleson, and Rivlin 1970; Mehrabian and Russell 1974; Moos and Insel 1974; Saarinen 1969; White 1973; Sims and Baumann 1974a). But a survey of this literature reveals several gaps. Most of the previous works have dealt with only one type of pollution such as air or noise pollution, and have not considered important elements of interaction among various pollutants. Also, the studies have either evaluated perception of pollution in relation to the social characteristics of the respondents, evaluated perception of pollution in terms of the actual levels of pollution, or merely conducted opinion surveys on whether people see pollution as a problem. None of the studies has applied attitude theory in an actual survey of community or individual attitudes.

Appendix 1 tabulates information about previous studies of attitudes toward pollution, including authors, site of the study, years covered, sample population, sample size, pollutant, type of questionnaire, concepts measured, relationship of social characteristics to attitudes, and relationship of exposure levels to attitudes).

Discussion of Sample Studies
Air Pollution

Several problems are evident in the previous studies of

attitudes toward air pollution, including the relatively small sizes of the areas sampled and the nonrepresentativeness of the sample populations (e.g., high-school students do not necessarily reflect the attitudes of the general population).

One of the earliest studies was by Smith, Zeidberg, and Schueneman (1964), who related levels of concern about air pollution in Nashville to both social characteristics and pollution levels. They were the first to develop an index of concern, based on whether a person was bothered by air pollution, with a checklist that indicated how he was bothered. A socioeconomic score was derived, using education, occupation, amount and source of income, condition of the dwelling, and number of persons per room. These scores were then grouped into high, low, and medium social standing.

Their findings indicated that pollution levels were not the only factor influencing awareness and concern, although there was a positive correlation and although concern for health and property damage from air pollution was positively related to concern about air pollution itself. But the statistical analysis of the data was weak and no statistically significant results were produced. Further, the survey was biased since 92 percent of the sample was female.

De Groot et al. (1966) examined levels of awareness, concern, and perception of the seriousness of air pollution in Buffalo, New York. They found that concern was not related to social characteristics, and that level of exposure was the key variable in determining concern about air pollution; there was a direct relationship between the quality of the ambient air and the perception of the seriousness of air pollution.

Medalia (1964) observed that attitudes toward air pollution were most significantly related to general satisfaction or dissatisfaction with the community, belief about the need for action to abate air pollution, and length of residence in the community. Lesser relationships were observed between attitude and occupational status, distance from point of emission, and interaction of gender with length of residence.

Crowe's (1968) study was unique in that it developed a definitional model of air pollution--respondents were asked to define the term. Their responses were then categorized into causal definitions (point sources), effectual definitions (dirty windows, paint deterioration), specific definitions (smoke), and combination definition. On the basis of this model and in comparison with social characteristics, Crowe found that more highly educated people saw pollution as a complex event stemming from a particular source and had little concern with effectual definitions. Place of residence had no significant bearing on the type of definition, and length of residence and sex also were insignificant. Finally, the higher the respondents' social standing, the greater the tendency for them to define air pollution in causal terms.

Finally, a number of studies have dealt with coping strategies or alternative adjustments (Kromm 1973; Kromm, Probald, and Wall 1973; Wall 1973 a, b; Unwin and Holtby 1974). The Unwin and Holtby and the Wall studies dealt with the effectiveness of smoke-control legislation in England. Wall found that respondents did not regard themselves as major contributors to air pollution either individually or collectively. Though they recognized the deleterious effects of pollution, they denied that it affected them financially. And though they acknowledged that air pollution was a problem in their home areas, they denied that they as individuals were adversely affected by it. The Kromm studies again dealt with adjustments to pollution and were modeled after the natural-hazards research studies; they used cross-cultural comparisons and systematized the research for future cross-cultural work.

Water Pollution

A large number of the water studies have concentrated not on pollution itself, but rather on watershed development, drinking water, renovated waste water, and water problems in general. Only two studies have related attitudes toward water pollution and levels of exposure (David 1971; Ditton and Goodale 1974).

Frederickson and Magnas (1968) compared city and suburban respondents in Syracuse, New York, on priorities of problems, attitudes about water pollution, what individuals or groups considered water pollution to be, and whether water pollution control was more or less important than other governmental responsibilities. Responses to these questions were the dependent variables in an analysis in which the independent variables were education, length of residence, age, and income. They found that the variance in attitude was due to differing personal characteristics. As socioeconomic status decreased in the city, so did level of concern. The higher the socioeconomic status and the lower the age, the higher on the scale of priorities water pollution was placed, and the more willing the respondents were to be taxed. The authors concluded that water pollution control was indeed a middle-class concern.

David (1971) examined the definition of water pollution and the indicators of pollution in Wisconsin, comparing standards and measures of water quality with perceptions of water quality. Again social variables were used to explain perception; family income, education, length of work week, age, sex, family composition, and size of residence. She found that poor water was mentioned most often by those who lived closest to it. Women were more likely to see pollution as a problem than were men; people living in small towns were more likely to see it as a problem than city dwellers; the presence of algae was felt to be the most important indicator of pollution.

The most comprehensive study yet completed on water pollution was done by Gore, Wilson, and Capener (1975) in upstate New York. They were specifically interested in the dimensions and sources of concern with abatement measures. Their study differentiated respondents by nearness to running water (streams) versus standing water (lakes). They found a negative weighting of perceived cost on willingness to support public action and willingness to act personally--people were less willing to act to control pollution when the cost was high. A negative weighting of water use and awareness of pollution led to the conclusion that the more water a person used the less aware she was of pollution. The perception of the role of government had a strong effect on willingness to support public action. Finally, no support was found for the direct effect of social or media participation or socioeconomic status on individuals' willingness to act.

Noise Pollution

Noise pollution studies have focused upon aircraft and air force noise, and on sonic booms as components of the problem, not on community noise levels. The earliest studies were mainly concerned with annoyance reactions, adaptability to the stimulus, tolerability, and the physiological and behavioral effects of noise.

The most important studies were the London noise study (Committee on the problem of noise 1963), and Bragdon's 1970 study of Philadelphia. The London study related levels of exposure to noise pollution and the perception of noise (annoyance scale). Residents within a ten-mile radius of Heathrow airport were sampled. The researchers were interested in concern, the effects of noise on the respondents' activities, the tolerance of the respondents, and their adaptability to high noise levels. When noise levels and annoyance scores were plotted for different frequencies of flyovers, the relationship was found to be linear and equal in slope, varying only at the intercept. A noise number index (NNI) was then developed. As the NNI increased, so did the percentage of the people who disliked the neighborhood because it was too noisy. There was no noticeable increase in the percentage of people who got used to the noise, nor was there any appreciable decrease. This study was the first comprehensive treatment of the subject and the relation between exposure levels and annoyance, tolerance and so forth.

Bragdon (1970) was interested in how people responded to community noise, the levels of exposure, and whether community noise constituted a problem of public concern. This was a very thorough study; the sample population accurately reflected all aspects of the total population with the minor exception that there were a larger proportion of women, and a large number of social variables were used. The rating of the neighborhood in terms of noise was used as the base variable for the social analysis. Bragdon found that homeowners were more concerned about noise than were apartment-dwellers and that residents with air conditioning in their homes perceived community noise levels as less of a problem than did those without air conditioning. Age and mobility did not affect noise perception at all, but the place where the respondent grew up (city, suburb, country) did. If the respondent was raised in an urban environment he was less critical of the community noise level. Finally, education, race, sex, and income were not correlated in any statistically significant way with perception of noise pollution.

Solid Waste, Pesticides, and Hazardous Chemicals

Klee (1971) looked at awareness of and attitudes toward sanitary landfill sitings. He was primarily interested in whether attitudes toward sanitary landfills varied with distance from the landfill. He found that they did.

Sigler (1973) and National Analysts, Inc. (1973) studied solid waste pollution. Sigler was interested in awareness of solid waste as a problem, attitudes toward solutions to solid waste and other environmental problems, and the effect of social variables in explaining these differences in attitudes. She found that there is a hierarchy among pollution problems for the people of Illinois, with pollution of rivers and lakes ranking highest, followed by air, solid waste, noise, and visual pollution. In relation to the seriousness of pollution and attitudes toward pollution, she found an inverse relationship between age and perceived seriousness of the problem (younger people felt the problems were more serious than did older persons); a positive relationship between education and perceived seriousness (more highly educated people rated pollution as more serious); the same positive relationship for income; and no relationship between perceived seriousness and sex; and, as she expected, the highest correlation was with city size. Sigler also found that the higher the socioeconomic status the more the respondents were willing to or currently did participate in recycling operations. This study is less statistically valid and significant than some of the others.

National Analysts (1973) were concerned with solid waste pollution in itself, not as a component of an overall survey, as was the case with Sigler's study. They looked at the level of knowledge of solid waste practices in the respondent's community; the influence of advertising and packaging on consumption habits; how respondents considered reducing the amount of solid waste in their homes; the acceptability and use of products made of recycled or reclaimed materials; and the current and potential interest in recovery operations.

They found little awareness of disposal practices among their housewife respondents, who did not know the cost of collection or disposal. Though they had little knowledge of local recycling activities, they were highly aware of solid waste as a problem. This study was helpful because it dealt solely with solid waste and solid waste management practices, but the bias of the sample (housewives) and the small sample size lead one to question the oeverall applicability of the results.

The most unusual study in this group involved litter. Finnie (1973) observed people buying hot dogs wrapped in wax paper from a street vendor and calculated the rate of littering as a function of individual characteristics and environmental variables. The environmental variables included a clean versus a dirty environment (litter or no litter on the street) and trash receptacle versus no trash receptacle in the immediate area. The social characteristics included age, sex, social status (white- or blue-collar), and race. Finnie found that litter receptacles substantially reduced littering. He also found that people were less likely to litter a clean area than a dirty area. People under eighteen had a higher littering rate than those over eighteen, and blacks had a higher rate than whites.

Multiple-Pollutant Studies

Nine studies, varying in quality, were found that dealt with more than one pollutant. The most significant were Van Arsdol, Sabagh, and Alexander (1964) and Jacoby (1972).

Van Arsdol, Sabagh, and Alexander, in one of the earliest studies on the topic, compared the perceived seriousness of a hazard with the actual exposure levels. They wished to determine the spatial aspects of hazards (defined in this study as smog, air-traffic noise, brush fires, earth slides, and floods) in the Los Angeles area and in learning how residents perceived the seriousness of these hazards.

The seriousness of the problem thus became the dependent variable and the independent variables were presence or absence of the hazard, socioeconomic status (measured by the Duncan-Reiss Index), family structure, type of dwelling unit, race, age, sex, and satisfaction with the neighborhood.

The authors found that there was indeed an association between the presence of a hazard and the perception of it as a problem. Sex and type of dwelling had no relation to perception. Age was a factor only in the case of landslides, where older persons were more likely to perceive the hazard. Race was a factor in smog and air-traffic noise perception, with whites more likely to perceive this as a problem. Finally, the most significant indicator of perception was socioeconomic status in the case of landslides. This is not surprising, given the spatial patterning of housing in the area. Those with higher incomes can afford to live in the hilly areas of Los Angeles where there is the greatest risk of landslides.

The conclusions found by Van Arsdol, Sabagh, and Alexander most relevant to the present study are: (1) perception of hazards (smog, air traffic noise, brush fires, and earth slides) was directly related to exposure level; (2) there seemed to be no consistent influence of social characteristics upon perception; (3) a respondent's expressed satisfaction with her neighborhood affected how she perceived the hazards.

The most comprehensive study on perception of pollution as it relates to social and environmental parameters was done in Detroit by Jacoby (1972). His study dealt only with air, noise, and water pollution, using concern about pollution as the dependent variable and the following as independent variables: level of pollution; socio-economic status; temporal variables (including length of time in Detroit and length of time in dwelling unit); spatial variables (place where the respondent grew up, perception of pollution as a neighborhood or regional problem versus as a national problem, and general quality of the respondent's physical milieu as measured by the quality of housing and pollution variables); and the importance accorded to specific pollution problems compared with other urban problems.

The relationship between exposure and concern was examined using correlation analysis. Concern about noise pollution was correlated with noise levels (dBA [Weighted Decibel] Scale) by a factor of +.23; the correlation coefficient of air pollution concern and air quality (measured by dustfall) was +.40; and the correlation coefficient of water pollution perception and water quality (measured by use of waterways for recreation) was +.65. As one can see, these correlations are not very high (excluding the case of water).

Noise pollution was considered a problem by the low-income, low-status portion of the population, contrary to other findings in the literature, and concern for noise pollution was greater in areas of lower housing quality. Higher-income whites were more likely to be concerned about air pollution if they lived in areas with higher pollution levels. This again is contrary to previous findings in the literature. Because of the methods of measuring water quality, the results of the study on this specific parameter are questionable. The overall conclusion of the study is that pollution levels are indeed related to levels of concern.

Jacoby concludes that concern about pollution is not limited to any particular social class or to any one income, age, or racial group. Increased exposure to pollution (intensity and duration) leads not to decreased concern but to increased concern, measured by personal encounter with pollution.

Summary of Research to Date

The principal conclusions of these studies can be summarized as follows:

1. There is a high awareness of pollution as a problem,

regardless of the pollutant.

2. Awareness of pollution and concern about it are related to exposure levels; exposure level is the key variable in determining concern about pollution (Smith, Zeidberg, and Schueneman 1964; De Groot et al. 1966; Bragdon 1970; Committee 1963; Jacoby 1972; Van Arsdol, Sabagh, and Alexander 1964; Medalia 1964; David 1971; Ditton and Goodale 1974).

3. Variance in attitudes is most related to satisfaction with the community (Medalia 1964; Van Arsdol, Sabagh, and Alexander 1964; Committee 1963; Bragdon 1970; Cooke and Saarinen 1971).

4. There are no consistent findings relating social characteristics and attitudes (Jacoby 1972; Van Arsdol, Sabagh, and Alexander 1964). Some studies conclude that concern is not related at all to social indicators (De Groot et al. 1966) and that these characteristics do not affect an individual's willingness to act to alleviate the problem (Gore, Wilson, and Capener 1975). Other studies conclude that specific social characteristics do influence perception (McKennell 1970; Kromm 1973) and attitudes toward pollution.

a. Length of residence was a factor in determining attitudes in three studies (Crowe 1968; Medalia 1964; Frederickson and Magnas 1968).

b. Age was also seen as a factor by Frederickson and Magnas (1968); David (1971); and Sigler (1973); as well as by Van Arsdol, Sabagh, and Alexander (1964) in the case of slides. Bragdon (1970) concluded that age was not a determining factor in the perception of noise.

c. Income and education were important in explaining perception variance in three studies (Sigler 1973; Frederickson and Magnas 1968; and David 1971).

d. Sex was important in one study (David 1971), as was race (Van Arsdol, Sabagh, and Alexander 1964) and mobility

(Bragdon 1970).

   e. Socioeconomic status on a general level appeared as a significant explanatory variable in perception of the problem of pollution only in the studies by Smith, Zeidberg, and Schueneman (1964), Crowe (1968), and Van Arsdol, Sabagh, and Alexander (1964)--the last for landslides only. Low socioeconomic status was equated with low levels of concern in David's (1971) and Frederickson and Magnas's (1968) studies. High socioeconomic status was correlated with placing water (David 1971) and solid waste (Sigler 1973) pollution high on the list of urban problems, as well as with the willingness to participate in recycling (Sigler 1973). On the other hand, low income and low status were significant indicators of high concern with noise pollution in areas of lower housing quality (Jacoby 1972). High-income whites were more concerned if they lived in high-pollution areas (Jacoby 1972).

   f. Homeowners exhibited more concern for noise pollution than did apartment-dwellers (Bragdon 1970).

   g. Finnie (1973), in his experimental study of litter, pointed out that litter is a function of social characteristics and environmental pollution levels; people under eighteen have a higher probability of littering, as do blacks, and people are less likely to litter a clean area than a dirty one.

   5. There is little awareness of solid-waste-disposal practices, cost of disposal, and local recycling efforts, although there is a high degree of awareness of solid waste as a problem (National Analysts 1973).

   6. Respondents in some studies felt that they were not the major contributors to pollution either individually or collectively, nor did they feel adversely affected by it, even though they acknowledged pollution as a problem (Wall 1973 a,b).

   7. Some studies related coping strategies to social

characteristics and briefly to exposure levels (Kromm 1973; Kromm, Probald, and Wall 1973; Wall 1973 a,b; Unwin and Holtby 1974), but no definite trends or factors appeared.

Problems and Limitations

As can be seen from the previous discussion, there is a lack of precise empirical work on attitudes toward pollution, and there are conceptual problems with the definitions of the terms used and with specifying exactly what was measured. Thus we find some unresolved questions regarding attitudes toward pollution and pollution abatement.

1. Do attitudes about environmental pollution vary by the social characteristics of the population of a large urban region? The previous studies either have not been statistically significant because of problems with sampling procedures, sampling size, or data analysis; have been biased with respect to the sampling population (e.g., high-school students); have not been done in large metropolitan areas where a significant number of people were studied; have been merely superficial opinion surveys; or have been conceptually weak and not based on attitude theory, with no clear definition of the concepts they were in fact measuring.

2. Can variance in attitude toward pollution be explained by the actual levels of exposure to pollution? That is, do high levels of exposure to pollution influence attitudes toward pollution so that people feel the problem is serious, and are they bothered by it? Here again, the previous studies have failed to give sufficiently detailed empirical definitions of pollution levels.

3. Are attitudes toward doing something about pollution (e.g., coping strategies) influenced by social characteristics or attitudes toward pollution?

4. Do coping strategies vary with varying levels of

exposure? Questions 3 and 4 and their accompanying relationships have not been adequately addressed in the literature dealing with attitudes toward environmental pollution.

These questions provide the main focus of this research. We turn next to a review of the literature on attitude theory, in particular the specific attitude theory used to structure the questionnaire used in this study and the subsequent analysis.

CHAPTER III

ATTITUDE THEORY

For many years social psychologists have been debating the merits of various theories of attitude acquisition and change; many theories have come into vogue and then been discarded. There is still little agreement about what an attitude is, how it is formed, how it is changed, and what role it plays in influencing behavior. It is therefore essential to give a general overview of the most significant and widely held theories of attitude acquisition and change.

Basic Terms Defined

In any discussion of attitude theory it is necessary to clarify one's terminology. In this instance four major concepts need definition: <u>attitude</u>, <u>belief</u>, <u>behavioral intention</u>, and <u>behavior</u>.

Attitude

Attitude can be most appropriately defined as an organized structure of ideas with both affective and cognitive components, which results in some behavioral intent. It is the affective component that distinguishes attitude from other concepts. Attitude may be conceptualized as the amount of affect for or against some object or situation (Thurstone 1931; Fishbein and Ajzen 1975). However, there is widespread disagreement on this concept in the psychological literature. The terms <u>opinion</u>, <u>satisfaction</u>, <u>prejudice</u>, <u>intention</u>, <u>value</u>, <u>belief</u>, and so forth, have all been used to measure this affective component. The result has been obvious: to confuse the meaning of both attitude and these other concepts. Distinctions among these terms have been suggested (Rokeach 1968; Triandis 1971), but the prevailing view is that these distinctions are not warranted at this time, since it has not been shown by research that these concepts obey different scientific laws or produce different results in experimental situations.

Attitude theorists have consistently come up with the following trilogy to explain or define attitude: affect, cognition, and conation. Affect refers to a person's feelings toward and evaluation of some person, object, issue, or event; cognition denotes the person's knowledge, opinions, beliefs, and thoughts; and conation refers to a person's behavioral intentions and actions.

In Fishbein and Ajzen's (1975) terminology (which I use), a distinction is necessary between behavioral intent and behavior. This suggests a classificatory system based on four broad categories: affect (feelings); cognition (opinions, beliefs); conation (behavioral intentions); and behavior (observed overt acts). Attitude will thus be used here for the first category; beliefs will be used for the cognition category; behavioral intentions will replace conation; and behavior will represent a fourth category.

Beliefs

Beliefs can be defined as the information a person has about an object. More distinctly, a belief links an object to some attribute. An example would be the statement "Pollution is harmful to health," where the object (pollution) is linked to the attribute (harmful to health). According to Fishbein and Ajzen (1975, p. 12), "the object of belief may be a person, a group of people, an institution, a behavior, a policy, an event, etc., and the associated attribute may be any object, trait, property, quality, characteristic, outcome or event."

Behavioral Intentions

The category of behavioral intentions refers to a person's intent to perform a specific behavior. It can be viewed as a special case of beliefs in which the object is always a person and the attribute is always a behavior.

Behavior

"Behavior" is self-explanatory; it means observable actions, which are generally studied in their own right. All questionnaire or verbal responses are instances of overt behavior, but they are often used to infer attitudes, beliefs, or intentions, as is done in this study.

In summarizing these concepts, Fishbein and Ajzen (1975, p. 13) point out that:

> . . . the concept "attitude" should be used only when there is strong evidence that the measure employed places an individual on a bipolar affective dimension. When the measure places the individual on a dimension of subjective probability relating an object to an attribute, the label "belief" should be applied. When the probability dimension links the person to a behavior, the concept "behavioral intention" should be used. Other concepts that have been employed in the attitude area appear to be subsumed under one or another of these three broad categories. For example, concepts like attraction, value, sentiment, valence, and utility all seem to imply bipolar evaluation and may thus be subsumed under the category of "attitude." Similarly, opinion, knowledge, information, stereotype, etc., may all be viewed as beliefs held by an individual.

The rest of this chapter summarizes the major theories of attitude acquisition, formation, and change, with a short evaluation and critique of each.

Theories of Attitude Acquisition and Change

Learning Theories

Some investigators have used the principles of learning theory (Hull 1943, 1951; Tolman 1932) to explain attitude acquisition and change. Two classical paradigms of conditioning--classical conditioning and operant or instrumental conditioning--are used to explain learning.

Most learning theories are concerned with how attitudes are acquired. In other words, how evaluative responses become associated with given stimulus objects. In some approaches to this problem, most notably States (1968), the paradigms of primary and higher-order classical conditioning are used. Persons who experience reward or reinforcement of some kind react with some observable goal response, which is considered synonymous with attitude.

One of the earliest applications of these learning principles to attitude theory was by Leonard Doob (1947). He viewed attitude as an unobservable response to an object that occurred before or in the absence of any overt response. Doob emphasized that a person first learns an implicit response to a given stimulus, then must learn to make the specific overt response to that stimulus. The first process (response to a given stimulus) can be accounted for by classical conditioning; the second process (overt response to attitude) can be accounted for by instrumental learning. According to Doob, this entire mediating response (both processes) constitutes the attitude. He distinguished between attitude and other types of responses by reserving the term "attitude" for those implicit responses that are elicited by socially relevant stimuli--that are socially significant to the person.

The reinforcement theory of attitude change has received its greatest input from Hovland, Janis, and Kelly (1953) and other researchers at Yale. Their theory is based on the initial learning theories of Hull as modified by Doob. In essence, the reinforcement theory states that attitude change results from learning through reinforcement. This group got its start during World War II and was involved in theories of persuasion and persuasive communications. After the war there was a growing interest in attitude acquisition and change, and most of the research was done in the laboratory.

Hovland, Janis, and Kelly (1953) maintain that one of the ways persuasive communications give rise to attitude change is by producing related opinion change. In their terminology, opinions are "beliefs such as interpretations, expectations, and anticipations." Attitudes, on the other hand, are implicit responses oriented to avoiding or approaching--reacting favorably or unfavorably toward--any object or symbol. Both opinion and attitude are regarded as intervening variables, of which there is a high degree of association and interaction. According to this theory the most important interaction is the change in attitude following a change in opinion.

The theory states that there are common principles that apply equally to the learning of new opinions and verbal responses and to the learning of motor skills. However, Hovland, Janis, and Kelly (1953) distinguish between learning or acquiring new opinions and learning new attitudes. When confronted with a persuasive communication the individual may react in two ways: he may think of his own answer to the question, or he may respond affirmatively to the answer suggested by the communication. The acceptance of the new opinion is contingent upon incentives that are offered by the communication, such as arguments supporting the conclusions or rewards or punishments that are expected to follow the acceptance of this new opinion.

According to Hovland, Janis, and Kelly (1953), there are three important variables in the acquisition of new opinions: attention, comprehension, and acceptance. Before any persuasion can take place, the individual must be attentive to the communication. Second, she must understand and comprehend its contents. Finally, in order for the communication to be accepted, significant incentives must be present. The theory does not attempt to catalog all the expectations that may influence acceptance, but it does point out three of the most important: expectation of being wrong or right; expectation of manipulative intent (a communication coming from an untrustworthy source is not likely to be accepted); and expectation of societal approval or disapproval.

Finally, the persistence of opinion over time depends upon the retention of the information content of the persuasive communication and the incentives for acceptance. Even if there is full retention of content and incentives, the long-range effectiveness of the communication will depend on how well it can resist subsequent conclusions from other communications.

The reinforcement theory of attitude change was intended to represent a preliminary statement or initial framework for future theory-building, and as such it is open to some major criticisms. First, it does not state connotatively, rather than just denotatively, what a reinforcing stimulus is. The three main variables--attention, comprehension, and acceptance--are the key concepts. But the connections between these concepts have not been thoroughly worked out by the authors, and the existing literature is vague and inconclusive about these relationships. As Insko (1967, p. 62) suggests, "What about the relationship between attention and acceptance? While attention may be in many cases a necessary prerequisite for acceptance, the possibility remains that many attitudes or opinions may be acquired or accepted incidentally without the aid of explicit attention. If so, how do such attitudes or opinions differ from those acquired through a process of attention?" This raises questions about the effect of varying amounts of attention on persuasive communications.

One of the main assertions of this theory is that attitude change occurs subsequent to opinion change. But the theory says very little about how this occurs or the reasons for it. Finally, there is very little explanation of how a persuasive communication supplies reinforcement for the advocated point of view. Arguments in the communication may be one set of incentives, but there is no elaboration of this idea, nor is there any evidence in the theory to confirm this as a major incentive mechanism.

Heider's Balance Theory

Balance theory can be placed under the general rubric of consistency theories or equilibrium theories of attitude acquisition and change. Balance theory as developed by Fritz Heider (1944, 1946, 1958) and modified by Newcomb (1953, 1959) is an attempt to understand the factors that link attitude toward an event to attitude toward the person causing the event. According to Heider (1946, p. 107), "Attitude towards persons and causal unit formations influence each other. An attitude towards an event can alter the attitude towards the person who caused the event, and if the attitudes towards a person and an event are similar, the event is easily ascribed to the person. A balanced configuration exists if the attitudes towards the parts of a causal unit are similar."

Heider theorizes about two types of relations between persons and between persons and events or things. A _sentiment_ relation is an attitudinal relation that implies admiring, liking, approving, and so forth. _Unit_ relations result in a perceived unity of persons or persons and events. Examples of unit relations are proximity, membership, causality, and ownership. In his symbolic terminology, Heider uses $\underline{p}$ as the focal person, $\underline{o}$ as another person, and $\underline{x}$ as the object or event. A negative sentiment relation would be represented by $\underline{nL}$, a positive sentiment relation by $\underline{L}$. A positive unit relation would be represented by $\underline{U}$, whereas a negative unit relation would be $\underline{nU}$. Therefore a negative sentiment relation with an object would be represented by pnLx. In his emphasis on unit formation and liking, Heider has developed triadic relationships between persons and between persons and things.

By a balanced state Heider means a state where everything fits together harmoniously without stress in the person's (p) life space. A lack of balance results in stress and pressure for change. In a more specific mode, balance refers to the tendency to make all the sentiment relations agree with each other--for example, to love those we admire. This tendency toward homogenous sentiment relations results in judging others as completely good or completely bad.

Heider also discusses balance as a relationship between parts. In dyadic relations involving p and x, balance exists if both the sentiment relation and the unit relation are of the same sign (either both positive or both negative). Symbolically this would be pLx + pUx or pnLx + pnUx. This same relation holds for the dyad between p and o. Balance theory, then, implies that interaction and proximity result in positive sentiment relations.

In triadic relations, balance exists if all three signs are positive or if two are negative and one is positive. Concerning three negative signs Heider is somewhat ambiguous, but he does say that this system of three negatives is unstable and tends toward a system of one positive and two negatives because of mutual dislikes. Heider prefers not to commit himself on the question of balance in this instance.

In Heider's theory imbalance results in tension that forces a shift toward balance, which can occur by change in the unit relation or the sentiment relation. An example of an imbalanced relation would be the following: p likes o, who has done something, x, of which p does not approve. To balance this, one could change the sentiment relation to p dislikes o, or o approves of x; or balance can be restored by changing the unit relation, so that p begins to think that o is not really responsible for x. In schematic form, balanced and imbalanced triads would be represented as in figure 1.

Although balance theory has wide applicability in attitude change, there are some limitations to it. First, Heider has dealt only with the qualitative relations between the entities, although subsequent work has led to some quantification. Second, the unit concept is generally vague. Two units are connected by a unit relation if the person perceives them as belonging together in some way or another, which leads to confusion concerning opposite and complement relations. The third criticism is that Heider deals only with the relations between a maximum of three entities. He does not discuss the degree of balance that exists in multiple relations, for example, complex dyadic or triadic configurations. Fourth, there is some question whether balance implies complete stability. Heider does not talk of degrees of like or dislike, although in real life there is such a variance. Fifth, he assigns both affective and connecting properties to the relations between cognitive entities. Sixth, there is no statement about which of the many ways to reduce imbalance will be taken in any given situation. Finally, the theory is unclear on what predictions are to be made when attractions among the triadic relations are of different origins and natures. Zajonc (1960) has proposed the following example: Suppose a person likes chickens, and chickens like chicken feed. The person thus ought to like chicken feed or suffer imbalance. This is a common problem that arises in logical reasoning when factual or truthful information is ignored. Logically this statement is correct, but factually it is not. In Zajonc's extreme example it has become necessary to distinguish between liking to eat chickens and other reasons for liking chickens.

Aside from the criticisms, balance theory is unique and

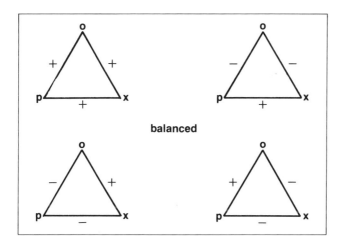

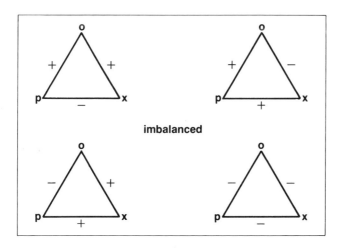

Fig. 1. Balanced and imbalanced triads. After Fishbein and Ajzen (1975). All triads reflect p's perspective. Lines between elements represent either unit or sentiment relations. The + and - signs represent positive and negative relations, either unit or sentiment.

rich in its applications. It not only deals with relations between beliefs and attitudes, but also has important implications for belief formation and relations among beliefs. As Fishbein and Ajzen point out, "since a person's behaviors may be represented as beliefs, the balance model suggests that a person's beliefs and attitudes may be influenced by his behavior" (1975, p. 35).

None of the extensions of Heider's balance model have attempted to quantify the unit or sentiment relations. However, using the balance model as a base, other consistency models, Osgood and Tannenbaum's congruity theory, and Rosenberg and Abelson's affective-cognitive consistency model have attempted to quantify beliefs and attitudes and also have attempted to deal with multiple beliefs about the attitude object.

Osgood and Tannenbaum's Congruity Theory

Congruity theory was developed by Osgood and Tannenbaum (1955) from prior work on the measurement of meaning using semantic space. The three-dimensional aspect of this space included activity (active-passive), evaluative (good-bad), and intensity (strong-weak) parameters. The method for locating terms connotatively (as distinguished from the denotative or lexical meanings of words) was called the semantic differential. Osgood and Tannenbaum found that an evaluative factor on the semantic differential scale was the most pronounced factor in defining meaning, owing to its favorable-unfavorable component. This led them to ask, Why does meaning shift its location in semantic space? Some theory was needed to explain this shift, and consequently the congruity principle was developed.

Osgood and Tannenbaum considered this evaluative component to be attitudinal because of its pro-con aspect. Attitudes, according to these authors, are conceived as one of many dimensions in semantic space. Attitude objects can be located on a number of semantic dimensions--for example, strong-weak--but it is this pro-con, favorable-unfavorable dimension that defines the evaluative nature of the attitude involved.

The theory of congruity merely says that when two attitude objects of differing evaluations are linked with an assertion each object tends to shift to a point of equilibrium or congruity. Attitudes can be favorable (+), unfavorable (-), or neutral (0). Assertions can be positive (associated, in the Osgood and Tannenbaum lexicon) or negative (dissociated). An example would be the statement "Industry likes weak pollution-control legislation." Here we have two attitude objects (industry and weak pollution-control legislation) linked by an associative assertion (likes).

One of the main aspects of congruity theory is the notion that attitudes tend toward maximum simplicity. In other words, attitudes move (change) toward maximum polarization, either positive or negative. This applies to attitudes considered singly. Thus the attitude object tends to be considered either all good or all bad. When two or more attitudes are linked by an assertion, there is movement toward congruity or equilibrium. But, the authors contend, this movement in itself is but another manifestation of this notion of simplicity.

When a state of incongruity (unequal polarization of objects) exists, the evaluations of the two objects will tend to change in the direction of congruity or equilibrium. Osgood and Tannenbaum have used the seven-point semantic differential scale (values range from +3 to -3) to locate attitude objects and have devised the congruity formula to calculate the change toward equilibrium. The pressure for change or congruity for two objects linked with an associative assertion is:

$$P_{oj_1} = d_{oj_2} - d_{oj_1}, \qquad (3.1)$$

$$P_{oj_2} = d_{oj_1} - d_{oj_2} \qquad (3.2)$$

where P_{oj_1} is total pressure toward congruity for the first object,

P_{oj_2} is the total pressure toward congruity for the second object,

and d_{oj} is the evaluative scale position of the objects

For two objects linked with dissociative assertions, the statements are identical to equations (3.1) and (3.2) except that they are preceded by a minus sign. The signs merely give the direction of the pressure. However, evaluation of both objects does not change equally. <u>The more polarized object changes proportionately less than the less polarized object</u>. Mathematically, this is stated:

$$AC_{oj_1} = \frac{|d_{oj_2}|}{|d_{oj_1}| + |d_{oj_2}|} P_{oj_1} \qquad (3.3)$$

$$AC_{oj_2} = \frac{d_{oj_1}}{|d_{oj_1}| + |d_{oj_2}|} P_{oj_2} \qquad (3.4)$$

where AC = attitude change in the first and second object, respectively,

and P = pressure toward congruity.

A few hypothetical examples will clarify the application of these formulas. Suppose the President (evaluated as +1) endorses national pollution-control standards (evaluated as -2). If the President is the first object of judgment and national pollution control standards the second, the preceding equations can be solved.

$$P_{oj_1} = d_{oj_2} - d_{oj_1} \qquad P_{oj_2} = d_{oj_1} = d_{oj_2}$$
$$= -2 \, (-[+1]) \qquad \qquad = +1 - (-2)$$
$$= -3 \qquad \qquad \qquad = +3 \, .$$

The total pressure toward congruity is -3 for the President and +3 for national pollution-control standards. This means that as a result of the associative link, the President will be evaluated less highly and the standards more highly. To complete solving the equation:

$$AC_{oj_1} = \frac{|d_{oj_2}|}{|d_{oj_1}| + |d_{oj_2}|} P_{oj_1} \qquad AC_{oj_2} = \frac{|d_{oj_1}|}{|d_{oj_1}| + |d_{oj_2}|} P_{oj_2}$$

$$= \frac{2}{1+2}(-3) \qquad\qquad\qquad = \frac{1}{1+2}(+3)$$

$$= -2 \qquad\qquad\qquad\qquad = +1 \ .$$

In schematic view this will be clearer. On this seven-point semantic scale the President is situated at +1 and pollution-control standards at -2. At the equilibrium position as calculated, the President must move two steps in the negative direction (or down) while the standards move one step in the positive direction (or up) in order for equilibrium to be achieved.

```
+3                              +3
+2                              +2
+1  x President                 +1  x
 0                               0  ↓
-1                              -1  o  (President and pollu-
-2  x pollution-control         -2  x  tion-control standards
-3     standards                -3     at equilibrium)
```

In this theory, Osgood and Tannenbaum realized that in many instances a given assertion may not be believed. They assumed that some degree of incredulity would exist when there is an imbalance (e.g., when two positively evaluated objects are dissociated or when one positively evaluated and one negatively evaluated object are associated); hence they introduced a correction for this incredulity factor. The values of the incredulity factor, $\underline{i}$, increase as the total pressure toward congruity, or the absolute value of $\underline{P}$ increases.

Finally, an assertion constant was added to the model to account for the finding that the object of assertion tends to change more than would normally be predicted by the principle.

The congruity theory as developed by Osgood and Tannenbaum is a good attempt to state mathematically some propositions that appear to have some basis in common sense. This makes it easier to test the theory and its implications and points out some of the less obvious implications of the theory. It also focuses on the idea of average and the notion of thresholds; that is, the attitude remains inactive until inconsistency is too large, then incongruity is resolved.

There are some major criticisms of this theory. One of these is the introduction of the incredulity and assertion factors. As Insko (1967, p. 139) points out: "Neither of these corrections follows from the congruity principle and both are simply introduced, in an attempt to patch up and make more reasonable some of the congruity implications." Another weakness of the theory is that assertion, a crucial concept, is not defined, and no clear distinction is made between associative and dissociative assertions.

One of the major criticisms of congruity theory is based on the idea of averaging versus summing, since summation effects do occur in the perception of complex stimulus objects. This summary notion is not incorporated in the theory and may lead to a revision of it (see the work of Fishbein and his associates, who utilize a summation model). Roger Brown (1962) has somewhat followed this idea and points out that congruity theory allows for change or resolution of incongruity only through changed evaluations in two of the associated objects of judgment. According to Brown, some resolution may result through a differential change in one of the objects.

Rosenberg's Affective-Cognitive Consistency Theory

This theory was first developed by Milton Rosenberg (1956), and a later version was developed with Robert Abelson (Abelson and Rosenberg 1958; Rosenberg and Abelson 1960). As a starting point for the theory, attitude is defined by the commonly used pro-con affect toward some object or class of objects. These affective sets are usually intertwined with a set of cognitions or beliefs, and as such Rosenberg prefers to consider attitudes as including both affective and cognitive components. He is primarily interested in cognitions that relate attitude objects to other objects of affective significance in some instrumental way. In other words, Rosenberg is interested in the importance of the cognition (whether it is a value, belief, goal, or attribute) and the certainty or strength of the attitude object's relation to that cognition. Thus, value importance is defined as the object's importance to the subject as a source of satisfaction, whereas the strength aspect (perceived instrumentality) is the subject's estimate of whether and to what extent the cognition would tend to be achieved or blocked by a certain policy. The perceived instrumentality is the linkage between a central object and a peripheral object. The central object can be referred to as the affective component of attitude, and the peripheral objects can be referred to as the cognitive elements of the attitude structure. This pairing of cognitive elements connected by a relation is called the cognitive unit.

According to the theory, a change in the affective component of the attitude structure will result in a change in the cognitive component and vice versa.

In the later version of the theory, Rosenberg and Abelson have refined their concepts. Cognitive elements are the basic entities of human thoughts, involving the representation of concrete and abstract things. Three broad classes of elements can be distinguished: actor, means, and ends. Actors are cognitively represented persons, groups, or institutions; means are actions or instrumentalities; ends are outcomes, goals, or values.

Of all the relations that could exist between cognitive elements, Rosenberg and Abelson theorize about four: positive, negative, null, and ambivalent--symbolized by p, n, o, and a. An example would be: positive relations such as likes, helps; negative relations such as dislikes, opposes; null relations such as indifferent to, does not affect; and ambivalent relations, which are combinations of positive and negative relations.

One technique for measuring this cognitive structure or unit is for the subject to rate each value (cognition) for the degree of satisfaction or dissatisfaction (value importance) and for the extent to which that cognition would be attained or blocked by the instrumental effects of the attitude object (perceived instrumentality). An index of the cognitive structure is obtained by determining the algebraic sum of the separate importance-instrumentality products. Algebraically, this is represented as:

$$A_o = \sum_{i=1}^{n} I_i V_i , \quad (3.5)$$

where I_i = instrumentality (the probability that o would lead to or block the attainment of a goal or value i)

V_i = value importance (the degree of satisfaction of dissatisfaction)

n = the number of goals or value states.

According to this theory, when affective and cognitive components of an attitude are mutually consistent the attitude is in a stable state; when they are inconsistent (to a degree that exceeds the individual's present tolerance for such inconsistency), the attitude is in an unstable state and will undergo spontaneous reorganizing activity until one of the following states is reached. The reorganizing will continue until affective-cognitive consistency is attained or until an irreconcilable inconsistency is placed beyond the range of active awareness of the subject. Restoration to consistency is the most common outcome when persons are involved in affective-cognitive inconsistency with regard to social attitudes.

Conceptual structures may be either balanced or unbalanced. Except for the rules involving the null and ambivalent cases, balance in Rosenberg's usage is identical to that in balance theory; any triad involving an even number of negative relations is balanced.

Inconsistency or imbalance can be resolved in three ways: change in one or more of the relations; redefinition or differentiation of one or more of the elements; and ignorance of the inconsistency or cessation of thinking about it. In some of their empirical work Rosenberg and Abelson maintain that the operation involving the fewest changes is most likely to occur. According to Abelson(1959), imbalance resolution has four operational modes: denial, bolstering, differentiation, and transcendence. Denial involves altering one or more of the cognitive elements. Bolstering merely blots out the inconsistency by bringing additional consistent relations to one or the other of the inconsistent cognitive units. In differentiation goals are lowered or redefined. And transcendence involves relating both inconsistent cognitive units to a larger superordinate concept or element. This is generally the last mode attempted because of the difficulty of creating a convincing superordinate structure.

Rosenberg and Abelson's affective-cognitive consistency theory is perhaps the most sophisticated statement of the consistency point of view in attitude theory. Basically built on Heider's balance theory, affective-cognitive consistency theory has been a definite improvement with its notion of the conceptual arena or cognitive structure and its theorizing about the imbalance resolution. But, like balance theory, this theory has failed to account for the varying strengths of relations.

The research on this theory is most consistent and supportive. Many of the aspects remain untested, but the evidence is encouraging for the usefulness of the formula and for the general adequacy of the formulation. As will be seen later in the discussion of value-expectancy models, there is a close parallel between Rosenberg's procedure of attitude prediction based on summing of products of value and instrumentality ratings and Fishbein's summation formula.

Festinger's Dissonance Theory

Dissonance theory, developed by Leon Festinger (1957), is another statement of the consistency point of view but is quite different from the consistency theories already discussed.

Festinger focuses on cognitive units--units are knowledge about various objects, facts, situations, circumstances, behaviors, and so forth. In the terminology I employed here, Festinger's cognitive elements would include beliefs, opinions, and attitudes.

The relationships between cognitive elements are of two types: irrelevant and relevant. In irrelevant relationships the two cognitive elements are completely unrelated (e.g., it rained last week, and the mail was delivered late today). Relevant relations can be of two types, dissonant and consonant. According to Festinger (1957, p. 13), "two elements are in a dissonant relation if, considering these two alone, the obverse of one element would follow from the other." In other words, x and y would be dissonant if not-x followed from y. Consonant relations imply that one cognitive element does follow from another.

The basic hypothesis of dissonance theory can be stated (1957, pp. 3, 18): "The existence of dissonance, being psychologically uncomfortable, will motivate the person to try to reduce the dissonance and achieve consonance. . . . The strength of the pressure to reduce the dissonance is a function of the magnitude of the dissonance."

The magnitude of the dissonance between two cognitive elements depends upon both the importance of the elements to the individual and the proportion of relevant elements that are dissonant. If the magnitude is great, there is more pressure to change or reduce the dissonance. Dissonance can be reduced in any of three ways: changing a behavioral cognitive element, adding new cognitive elements, or changing an environmental cognitive element. Examples are the following (Insko 1967, p. 199):

> The changing of a behavioral cognitive element is illustrated by the smoker who stops smoking when he learns that smoking is detrimental to health. Changing an environmental cognitive element is illustrated by the person who cuts a hole in his living-room floor to make more reasonable the fact that he always jumps over that spot, or who distorts the perceived political orientation of a candidate in order to justify the fact that he has voted for him. Adding new cognitive elements is illustrated by the smoker who reads material critical of the research linking smoking to lung cancer or material on the high death rate resulting from automobile accidents.

The maximum dissonance that can possibly exist between any two elements is equal to the total resistance to change of the less-resistant element (Festinger 1957, p. 28). When the less-resistant element changes, the magnitude of the dissonance is consequently reduced.

The magnitude of dissonance is one of the key concepts of this theory. In a formal presentation, this can be seen as:

$$D_k = \frac{\sum_{d=1}^{n} I_d}{\sum_{d=1}^{n} I_d + \sum_{c=1}^{m} I_c} , \qquad (3.6)$$

where D_k = magnitude of dissonance associated with cognitive element k

I_d = importance of dissonant element d

I_c = importance of consonant element c

n = number of cognitive elements in a dissonant relation with element k

m = number of cognitive elements in a consonant relation with element k.

Brehm and Cohen (1962), on the other hand, prefer to view dissonance as a function of the ratio between strength of cognitions of forces opposing the counterattitudinal act and strength of cognitions of forces supporting the counterattitudinal act. The closer to 1 the ratio becomes, the more dissonant is the situation.

Festinger discusses four situations that give rise to cognitive dissonance--decision-making, forced compliance, voluntary or involuntary exposure to dissonant information, and disagreement with other persons. Decision-making results in dissonance when the person is faced with a choice between a number of alternatives and hence is in a conflict situation. He assumes that his choice has some unfavorable aspects and that the unchosen alternative has some favorable aspects. The theory predicts that dissonance may be reduced by increasing one's positive evaluation of his chosen alternative or by decreasing one's positive evaluation of the unchosen alternative. Dissonance may also be reduced by revoking the decision or by establishing some cognitive overlap or similarity between the alternatives (a compromise solution). The amount of change will be related to the magnitude of the dissonance involved. For example, a choice between two books would arouse less dissonance than a choice between a book and a movie.

In forced compliance a person is compelled to perform a behavior by public pressure before or without accompanying change in his private opinion. Compliance can be forced by the threat of punishment or by the prospect of reward for compliance. Once the person complies, dissonance results between opinion and behavior. The magnitude of the dissonance in this situation is a function of the amount of reward or punishment offered and the importance of the opinion or behavior involved. Dissonance that results from forced compliance can be reduced in three ways: reducing the importance of the opinion or behavior involved; changing the private opinion to agree with the public opinion; or magnifying the reward or punishment used to induce the behavior.

Involuntary exposure to new information either through mass media or through interaction with new people is potentially dissonance-arousing. If this new information is contrary to the cognitions already held, dissonance will result. Dissonance may be reduced by defensively misperceiving the new information, avoiding the new information, or changing one's opinion. Voluntary exposure to new information results from the desire to obtain additional data for future action or from the pressure to reduce dissonance after a behavior has occured.

Finally, agreeing with other people reduces dissonance, and disagreeing with others tends to increase dissonance. The magnitude of the dissonance resulting from disagreement is a function of the testing of the point of argument by empirical observation; the actual number of people agreeing or disagreeing; the importance of the issue in dispute; the attractiveness and the credibility of the disagreeing person; and the general extent of the disagreement. The greater the number of people who agree, the less dissonance is produced by someone else who disagrees. Dissonance resulting from disagreement can be reduced in three ways: changing one's opinion to agree with the dissenting person; persuading the disagreeing person to change her opinion; or increasing the perceived difference or undesirability of the disagreeing person (e.g., he is stupid or boring).

A great deal of research has been done on dissonance theory, both methodological and empirical. Much of the appeal of this theory is related to its ability to generate nonobvious predictions that can be empirically supported. For example, according to the theory, if a person chooses to listen to a disliked communicator, the greater the dislike the greater the influence.

There are several weaknesses to this theory. The first has to do with how dissonance is defined. Succeeding clarifications have resulted in an intuitive feeling about dissonance but not a precise definition. Other weaknesses are related to the roles of commitment and volition, the relations of pre- and postdecisional processes, and the course of dissonance reduction over time (Insko 1967, p. 283). A large body of empirical literature has developed to support dissonance theory, but there are some major flaws in the research. One of the main limitations is that subjects have been nonrandomly eliminated in some experiments for supposedly theoretical reasons (e.g., they did not reach a level of dissonance). Second, the experiments rely heavily on deception. Finally, suspicion, doubt, or some other type of arousal that has been found in some subjects studied might have hampered the experiments and biased the results.

Expectancy Value Theory

Thus far the discussion has centered on the various types of theories of attitude formation and change. It is possible to categorize these into two major approaches: behavioral and consistency. Under the behavioral approach are the learning theories and the value expectancy models. Under the consistency approach are balance theory, congruity theory, dissonance theory, and affective-cognitive consistency theory.

One implication of the behavior theory approach to attitude formation is that belief-formation follows the laws of learning. Whenever a belief is formed, some implicit evaluation associated with the response becomes conditioned to the stimulus object. This implicit evaluation constitutes an attitude and may have been formed as the result of prior conditioning. This conditioning paradigm implies that the attitude toward the object is related to beliefs about the object. This has been made an explicit part of Fishbein's theory of attitude formation. It can be described in the following way (Fishbein and Ajzen 1975, p. 29):

> (1) An individual holds many beliefs about a given object; i.e., the object may be seen as related to various attributes, such as other objects, characteristics, goals, etc. (2) Associated with each of the attributes is an implicit evaluative response, i.e., an attitude. (3) Through conditioning, the evaluative responses are associated with the attitude object. (4) The conditioned evaluative responses summate, and thus (5) on future occasions the attitude object will elicit this summated evaluative response, i.e., the overall attitude.

According to the theory a person's attitude toward an object is a function of his beliefs about the object and the evaluative responses associated with those beliefs. Operationally, this relationship between an individual's beliefs about an object and his attitude toward the object can be expressed as:

$$A_o = \sum_{i=1}^{n} b_i e_i \quad , \tag{3.7}$$

where A_o = attitude toward some object, o

b_i = belief i about o (e.g., the subjective probability that o is related to attribute i)

e_i = evaluation of attribute i

n = number of beliefs .

Before we get into a discussion of Fishbein's model (Fishbein 1967a, b; Fishbein and Ajzen 1975), let us discuss some of the other theories that have arrived at similar formulations in attempting to account for overt behavior. Most of the theories were based on the work of Tolman (1932), who felt that people learn expectations--that is, beliefs that a given response will be followed by some event. Since these events could be either positive or negative reinforcers, his argument boiled down to the idea that people would learn to perform a behavior they were expected to perform and that this would lead to positively evaluated events.

The best-known expectancy value model is the subjective expected utility model (SEU) of behavioral decision theory, first developed by Edwards (1954). According to this theory, when a person has to make a choice, he will choose the alternative most likely to lead to the most favorable outcome--the alternative that has the highest subjective expected utility. This can be represented by the following equation:

$$SEU = \sum_{i=1}^{n} SP_i U_i , \qquad (3.8)$$

where SEU = subjective expected utility associated with a given alternative

SP_i = subjective probability that the choice of this alternative will lead to some outcome, i

U_i = subjective value or utility of outcome i

n = number of relevant outcomes (since n usually refers to a mutually exclusive and exhaustive set of outcomes SP = 1.00 in most behavioral decision theory analyses).

In this model a direct link is assumed between SEU and behavior. One can see that the equation as formulated by Edwards is very similar to Fishbein's. In Fishbein's terminology, the SEU model deals with beliefs about the consequences of performing a given behavior ($SP_i \approx b_i$) and with the evaluations associated with the different outcomes ($U_i \approx e_i$). This SEU equation could be rewritten thus in Fishbein's terminology (remember that this is attitude toward a behavior, A_B):

$$A_B = \sum_{i=1}^{n} b_i e_i \ . \qquad (3.9)$$

Earlier in this chapter we discussed Rosenberg's affective-cognitive consistency model. In his formulation, the more a given object (an action or a policy) is instrumental in obtaining positively valued goals, and in blocking negatively valued outcomes, the more favorable the person's attitude toward the object. This formulation was influenced by the functional approach to attitudes, whereas the SEU model and Fishbein's were both developed within the framework of behavior theory. However, both the SEU model and Rosenberg's instrumentality value model deal with beliefs about the object and with associated evaluations or values. The functional approach to attitudes suggests that attitude formation and change can be understood only in terms of the functions the attitude holds for the individual. Attitudes are necessary because they permit the individual to achieve certain goal states (e.g., organize knowledge, express his views). In his early formulation Rosenberg was concerned with the extent to which an object aids or hinders goal attainment. In his later work he developed the theory of affective-cognitive consistency to account for the relation between beliefs and attitudes. Fishbein utilized the conditioning process to account for the same relations.

Although the two models are very similar, I have chosen to use Fishbein's for two reasons. First, it implicitly deals with behavior in trying to predict behavioral intent and in being firmly based in the behavioral point of view. Second, there is a large body of theoretical and methodological literature on the formation of this model as well as some application of the model by Fishbein and his associates.

Fishbein's Model for the Prediction of Intention

Fishbein's model is an adaptation of Dulany's (1961, 1968) theory of propositional control, which deals with the determinants of behavioral intentions. Dulany's interest in determining behavioral intentions grew out of his work on the role of awareness in studies of verbal conditioning. Fishbein's model, like Dulany's, is concerned with predicting behavioral intentions that are assumed to mediate overt behavior. The greatest asset of the Fishbein model is its simplicity. The original formulation of the model (1967) is:

$$B \approx BI = (A\text{-act})w_0 + (NB_p)w_1 + ([NB_s][MC_s])w_2 , \qquad (3.10)$$

where   B = overt behavior

BI = bheavioral intentions

A-act = attitude toward the behavior in a given situation

NB_p = personal normative beliefs

NB_s = societal normative beliefs

MC_s = motivation to comply with normative beliefs

w_0, w_1, w_2 = empirically determined weights.

There are two major components to this model, a personal or attitudinal component and a social or normative component. Upon further refinement of the model and more empirical testing, a new version was developed:

$$B \approx I = (A_B)w_1 + (SN)w_2 , \qquad (3.11)$$

where   B = behavior

I = intention to perform behavior, B

A_B = attitude toward performing behavior, B

SN = subjective norm

w_1, w_2 = empirically determined weights.

This simplified model is based on the following findings by Fishbein and his associates (Ajzen and Fishbein 1970, pp. 467-68):

> it was suggested . . . that the normative component also included the individual's personal normative beliefs (NB_p); i.e., his own beliefs as to what he should do in a given situation. It appears, however, that in many situations NB_p may serve mainly as an alternative measure of behavioral intentions. . . . One other point needs to be mentioned regarding motivation to comply (MC). The measurement of this variable has repeatedly proved to be unsatisfactory. . . . Research to date has indicated relatively little variance in this measure, and thus the results obtained with normative beliefs alone were as good or better than those obtained when NB was multiplied by MC.

Now let us discuss the components of the model. The first is the attitudinal component. Fishbein has generalized this by suggesting that it refers to a subject's beliefs about performing the act. In contrast to traditional approaches to attitude, the attitude in question in Fishbein's model is the attitude toward performing the given behavior (eq. 3.9), not the attitude toward the object or target of the behavior (eq. 3.7).

The second component of the model deals with the influence of the social environment upon behavior. The subjective norm is the person's perception that most people who are important to her think she should or should not perform the behavior in question. This is determined by the perceived expectations of specific referent individuals or groups and by the person's motivation to comply with these expectations. This is symbolized by:

$$SN = \sum_{i=1}^{n} b_i m_i \quad , \tag{3.12}$$

where b_i = normative belief (the person's belief that reference group or individual i thinks he should or should not perform behavior B)

   m = motivation to comply with referent i

   n = number of relevant referents.

There has been some question about the meaning of motivation to comply. This variable is best conceived as the person's general tendency to accept the directives of a given reference group or individual (Fishbein and Ajzen 1975, p. 306). Motivation to comply can be interpreted as the person's intention to comply with the referent in question. The determinants of this intention then are the same determinants discussed earlier with respect to any behavioral intention. The following equation can express this:

$$m \approx I_c = (A_c)w_1 + (SN_c)w_2 , \qquad (3.13)$$

where m = motivation to comply with referent R

   I_c = intention to comply with referent R

   A_c = attitude toward complying with referent R

   SN_c = subjective norm concerning compliance with referent R.

As indicated in equation 3.11, both components of the model are given empirical weights proportional to their relative importance in predicting behavioral intentions. The empirical weights are expected to vary with the kind of behavior being predicted. For example, normative considerations may be more important (e.g., expectations of friends) than the expected outcomes of the act (attitudinal considerations). In other instances the opposite may be true. In terms of the availability of the weights, Fishbein and Ajzen state (1975, p. 303):

> Ideally, the weights for the attitudinal and normative components would be available for each individual with respect to each behavior in a given situation. Since adequate estimates of this kind are not presently available, the practice has been to use multiple regression techniques, and standardized regression coefficients

have served as estimates of the weights for the theory's components. The present version of the theory, then, is a multiple regression equation where there are two predictors, A_B and SN, and the criterion is I, the behavioral intention under consideration.

Over the past few years Fishbein and his associates have conducted many investigations based on this model. These have included controlled experimental situations as well as application to specific social attitudes and intentions (e.g., buying certain products, signing up for alcoholic treatment programs, using contraceptives, cheating on exams, taking part in family planning, and engaging in premarital sexual intercourse).

Theory Evaluation and Rationale

After consideration I concluded that Fishbein's theory of behavioral intent was the most appropriate theory to use with my particular topic and focus. Fishbein and his associates have in fact tried the theory in real-world situations, with good results. The theory was easy to apply to a questionnaire format, which was perhaps the most important criterion for my purpose. Also, most of the other theories focus on attitudes toward some object. Fishbein's theory focuses on attitudes toward an intent to perform a specific behavior. So in the long run this theory is interested in the prediction of behavior, whereas the others are interested in the acquisition of attitudes and the mechanisms of attitude change. Since my primary purpose was to examine how people cope with or react to a variety of pollution-abatement strategies, Fishbein's model was the obvious choice.

Content of Previous Attitude Studies in Light of Attitude Theory

As we saw in chapter 2, there is a considerable body of literature on attitudes toward pollution. However, there appears to be one major flaw in all the studies: Are they really measuring attitudes? From a theoretical viewpoint it appears that they are not. They merely focus on a set of questions regarding either an opinion about a situation or a problem, and some belief about or evaluation of how bad the problem is. There has been no systematic attempt to relate attitude theory to the question of pollution, let alone to strategies for pollution abatement. This is the primary contribution of the present research: to understand attitude theory and apply it to real-world situations in order to get a better measure of community or individual attitudes; and to show that this is the most expedient means of delimiting community attitudes and responses to the various environmental planning programs and methods of pollution abatement suggested by local, state, and federal authorities.

The application of the model to the data collected by the questionnaire will be discussed in more detail in chapter 4, which focuses on the relevance of the process of investigation, questionnaire design, and applicability.

CHAPTER IV

APPLICATION OF THEORY: SAMPLE DESIGN, QUESTION-
NAIRE CONSTRUCTION, AND ADMINISTRATION
OF THE QUESTIONNAIRE

The questionnaire on community attitudes was designed in accordance with two main criteria; first, it had to be short enough to be administered by telephone; second, it had to elicit responses not only on attitudes toward a subject--pollution-- but on attitudes toward a behavior--abating pollution. The questionnaire was developed on the basis of the theory of behavioral intent; that is, questions related to a person's attitude toward an object or a situation, his evaluation of that situation, how important that object was on his scale of priori- ties, and to some measure of overt behavior (what he had actually done about abating pollution). Thus all the components of the theory were accounted for. We also hoped that a few open-ended questions would make the respondent feel free to volunteer any relevant information on her opinion toward pollution or pollution abatement.

Questionnaire Construction

Examining questionnaires from the literature (Jacoby 1972; Medalia and Finkner 1965; National Analysts 1973) gave an idea of the types and formats of questions that have been used in studying this particular topic. Robinson and Shaver (1973) were very useful for the psychological aspects of the question- naire, particularly affect measurement.

Initially the interview was set at eight to ten minutes

because of financial and time limitations. We also felt that a short questionnaire would provide a higher response rate, since people would be willing to spend ten minutes to answer it. The telephone was chosen as the medium because responses are instant and response rates are comparatively high. Telephone interviewing has been shown to be the most effective way to reach a large number of people in a short time.

The questionnaire opened with a question on how often the respondent worried about a list of ten common urban problems (Appendix 2, question 1), in order to measure the centrality of pollution as a problem relative to other social worries. The problems named ranged from crime to quality of education to cost of housing. Respondents were asked how frequently they worried about each: very often, from time to time, or not much at all. For each problem a "don't know/no response" category was included. There was no change in this question after the pretest.

Next was a measure of the intensity of the problem of pollution. Since knowledge of an object plays a part in one's beliefs and attitudes, it was felt that a standardized definition of pollution would help equalize these differences in knowledge. Pollution was defined by the interviewer as follows: "By pollution I mean air pollution, water pollution, noise pollution (which is annoying sounds such as traffic noise or jackhammers), and street garbage pollution. By street garbage pollution I mean such things as cans, paper, bottles, litter and construction junk which you find around outside your home or on the street." The respondents were also told at this point that the survey was designed to measure their attitudes toward pollution. The intensity question (question 2) asked the respondents to evaluate the pollution problem in their area as very serious, somewhat serious, or not very serious. Again, a "don't know/no response" category was included. Each respondent was then asked the seriousness of each of the four specific types of pollution in his area. This question also was not altered after the pretest.

To make the survey correspond to the theory of behavioral intent, some measure of social norms was needed; this was fulfilled by question 3 which basically asked if the respondent felt her neighbors worried more about pollution than she did, less than she did, or about the same. This gave some measure of societal pressure with regard to worry about pollution and the respondent's relative position in this worry scale. Also, a measure of affect--or how pollution emotionally affects respondents--was incorporated into the questionnaire. The original list of four emotions was narrowed down to three: mad (angry), annoyed, and depressed. The emotion response sick (makes you feel bad) was dropped because of the double meaning of health and general annoyance. Robinson and Shaver (1973) supply a more detailed discussion on the ranges of emotive responses. This question (question 4) and question 3 were not changed after the pretest.

The next major step in organizing the questionnaire was to develop the measures of attitudes toward pollution abatement or coping responses toward pollution. Originally a list of twenty-two actions was provided, which included active, or group, actions like joining groups or writing letters to congressmen and passive, or individual, responses like staying indoors or keeping the windows closed. Each respondent was asked whether he would consider taking this specific action and how effective he thought such action might be (provided he said he would take it). The effectiveness scale ranged from very effective to somewhat effective to not very effective. Again, a "don't know/ no response" category was included. After the pretest this list was condensed and some actions were combined (such as "join a neighborhood group concerned with environmental problems" and

",join an environmental group") to produce a more manageable list of fourteen. The question as finally used is question 5.

Finally, to comply with the theory some measure of overt behavior was needed. Initially a list of actions was provided and each respondent was aksed if she ever did any of them and if so how frequently (frequently, once in a while, once). The list of actions was a scaled-down version of the intent list. During the pretest the questionnaire was found to be redundant and tedious; this question proved the most bothersome to the respondents and also increased the administration time of the questionnaire to about thirty minutes. The question was therefore changed to an open-ended one that simply asked whether the respondent had ever taken any of the actions mentioned above (question 6) and if so which ones (question 6a). This also allowed the respondent to comment on any other type of action he took with regard to pollution abatement.

As a corollary of this we were also interested in why people did not do more to abate pollution, and this was the next question (question 7). Four main reasons were provided: too many hassles, not enough time, wouldn't do any good, and not really interested. These seemed likely to cover most of the responses that would be encountered, and this supposition was proved correct by the pretest. The interviewer was simply to circle all the reasons that applied to the respondent.

Thus the general format was set; a centrality measure, an intensity measure, social norms, affect, behavioral intent and evaluation, and overt behavior. We decided to look not only at attitudes toward pollution in general but also at attitudes toward a specific type of pollution as a comparison measure. We decided on solid waste pollution for a number of reasons: (1) we had the most comprehensive information on actual levels of solid waste in Chicago for the comparison of attitudes and levels of pollution; (2) the actions one could take to abate or avoid the problem were of both an individual and a group nature; (3) it was easily enough understood to be either a problem or not a problem; and (4) the literature review revealed that very little work has been done on people's attitudes toward this type of pollution and its abatement.

The same format was followed as in the general pollution case. The list of intents was narrowed to ten and included such things as paying refundable deposits on beverage containers and asking people to clean up after their dogs. Each respondent was asked whether she would consider taking the action, and if so how effective she thought it would be in abating solid waste pollution (or street litter). This question remained unchanged after the pretest (question 8). This was followed by a question on whether respondents had ever done anything to curb solid waste pollution and, if so, what they had done (questions 9 and 9a). This question posed the same problem of repetition as in the general pollution section, and so its format was also changed to open-ended. Following these two questions was a question on why people did not do more to abate solid waste pollution. The question is very similar to question 7, but the format was changed slightly, although the four possible reasons remained the same, because we wished to obtain specific information on the most important reasons for lack of action to abate solid waste pollution. We deviated from the general pollution part of the questionnaire here by asking whether there were any other reasons for not doing more to abate solid waste pollution. We felt that some aspects such as lack of understanding of the problem, not feeling it was a problem, and so forth, would be revealed by this open-ended question and that we might thus get a better idea of the relation between intent to abate solid waste pollution and actual behavior. It also provided a forum for the respondent to interject any other thoughts on the subject of pollution.

The final question on the survey (question 12) asked the respondent whether he would be willing to discuss his attitude toward pollution in more detail in the near future. This was included to determine who might be available for in-depth interviews at some later time and also to gauge how interested people were in talking about their attitudes toward pollution.

From Theory to Practice

Attitude, intent, and behavioral measures were calculated from the questions in the interview, based on Fishbein's theory of behavioral intent as discussed in chapter 3.

Attitudes toward Pollution (AO)

Attitude toward pollution was determined by the summation model, using nine variables. These included questions 1f, 2a, 2b, 2c, 2d, 2e, 4a, 4b, and 4c (see Appendix 2). The b_i component was the amount of favorableness or unfavorableness toward that belief. This probability measure was calculated by using question 1f. If the individual responded with very often, it was coded as +4; from time to time was coded a +3; for don't know/no response (basically a neutral answer), the code was +2; and not much at all was coded +1. The e_i component (the evaluation of the belief) was determined by scaling the response categories in each of the remaining variables. The questionnaire responses were recoded to reflect a bipolar scale, negative to positive. For questions 2a through 2e the coding went like this:

> very serious was coded as +2
> somewhat serious was coded as +1
> not very serious was coded as -1
> don't know/no response was coded as 0.

For questions 4a through 4c the recoding was this:

> yes became +1
> no became -1
> don't know/no response became 0.

An example of the $b_i e_i$ computing for question 2a is the following:

The individual responded that she very often worried about pollution, hence b_i equals +4.

This same individual responded that she thought pollution was very serious in her area, hence e_i equals +2.

$$b_i \times e_i = 4 \times 2 = 8 \ .$$

This same procedure was carried out for each of the eight variables (questions). These were then summed and the AO score produced for the general pollution case. Attitudes toward air pollution (AOA), water pollution (AOW), noise pollution (AON), and solid waste pollution (AOSW) were calculated in the same manner but using five variables. The variable seriousness of pollution was replaced by the specific pollutant variable. For example, attitude toward air pollution was calculated using questions 1f, 2c, and 4a through 4c.

Attitudes toward Pollution Abatement (AB)

AB was calculated in basically the same manner as AO with the $b_i e_i$ product. The variables included questions 5a through 5n on the questionnaire. The e_i component (in this case the evauulation of the outcome i, pollution abatement) was calculated as the yes/no/don't know response for each intention (e.g., yes = +2, no = +1, don't know = +1.5). This was then multiplied by the effectiveness scale or the second part of the question. This is the composition of the b_i component of the model, which is the belief that the behavior in question will result in outcome i (pollution abatement). Again, we used a bipolar scale with values ranging from -4 to +4. If a respondent answered no on the intention, the effectiveness was computed as -4. It was assumed that if a person did not intend to perform a given behavior his evaluation of the behavior's effectiveness in abating pollution would be negative. Since the positive range

of the scale went to +4, the -4 was used to provide a balance.
If the respondent answered the question with don't know, the
effectiveness was computed as 0. If the respondent answered
yes, the effectiveness score ranged from +1 to +4 depending on
her response to the question (i.e., if she thought the action
would be very effective the score was +4; if somewhat effective,
then +3; if not very effective, then +1; and if don't know, +2).
Example: For the intention to complain to authorities, a
respondent answered affirmatively to the question (+2) and
stated that he thought this action would be very effective
(+4); the $b_i e_i$ would then be computed thus:

$$2 \times 4 = 8 \; .$$

If he answered negatively, the $b_i e_i$ would be computed as +1 × -4
or, -4. A neutral or no-response answer would be +1.5 × 0, or
0. The $b_i e_i$ product for each of the intentions (all fourteen
of them) was then summed to give the AB score. This same
procedure was followed in calculating attitudes toward solid
waste pollution abatement (BSW), but using questions 8a through
8j. Attitudes toward air pollution abatement were calculated
using question 5, parts a, b, d, e, f, g, i, j, k, l, m, and n.
Water pollution abatement attitudes (ABW) used question 5, parts
a, c, d, f, g, h, j, l, and n, and attitude toward noise pollution
abatement (ABN) used question 5, parts a, d, f, g, i, j, k, l,
and n.

Influence of the Social Environment on Behavioral Intent or Social Norms (SN)

As was seen previously, SN is also a summative model. The
B_j in this instance is the normative belief (e.g., a person's
belief that a reference group i thinks she should or should not
perform behavior B), and m_j is the motivation to comply with the
referent group. For our purposes the referent group was taken

to mean society in general. The computation of the b_j was the yes/no/don't know response on the intent question (question 5, parts a-n); yes was coded +1, no was coded -1, and don't know was coded 0. The m_j measure or motivation to comply was computed using question 3. The response "more than you" was coded as -1; "less than you" was coded as +2; "the same as you" became +1; and "don't know/no response" was coded as 0. The peer variable and the yes/no/don't know responses on the intent variables were then multiplied and summed to produce SN. Example: the response to question 5a is multiplied by the response to question 3: Intent to perform behavior a = yes (+1) × the same as you (+1) = 1. This procedure was followed for each of the specific pollutants, with the relevant intent measures used in calculating the AB scores for the individual pollutant.

Intention to Perform the Behavior in Question (I)

The variable I was computed as the number of yes responses on question 5, parts a-n (How many of the actions were they willing to perform to abate pollution?). This procedure was followed for the intent to abate air pollution (IA), intent to abate water pollution (IW), intent to abate noise pollution (IN), and intent to abate solid waste pollution (ISW), using the relevant parts of question 5 for the first three and question 8 for intent to abate solid waste pollution.

Actual Behavior (B and BSW)

This variable was the number of behaviors the respondent had actually performed in an attempt to abate general pollution (question 6a) and solid waste pollution (question 9a).

Sample Design

The area selected for study was the city of Chicago. Three types of information were obtained: social differentiation, environmental pollution, and attitude measurements. The attitudinal information, gathered by a telephone questionnaire, is discussed in more detail later in this chapter. The questionnaire was based on a stratified sample of communities within the city and tapped random residences within these communities. The specific areas sampled are discussed in Appendix 3. Since the city contains 1,157,030 households a sample of 1,067 was needed to achieve a plus or minus 3 percent error at the 95 percent confidence level.

To force a comparison of the three sets of variables, it was necessary to examine the extremes of environmental pollution and social differentiation. This enabled us to see the contrasts between areas of unusually high and low pollution, as well as the extremes of the social scale--for example, high and low socioeconomic status, and black and white. On a larger scale this forced comparison of the extremes would make the results more viable. Hence, in the sample design, areas of high and low pollution (as defined by a preliminary pollution index) and of high and low socioeconomic status (SES) were selected for study.

A preliminary environmental pollution index was compiled using thirty-eight environmental parameters (components of air pollution, trace metals in the air, and solid waste pollution; see Appendix 4). Water pollution variables were not used, since they could not be analyzed by square mile. Noise variables were also excluded because there were large sections of the city with no noise data. Each square mile of the city was examined to determine whether a threshold level of pollution was exceeded there; if so, that square mile was identified as having a specific pollution problem. This process was repeated for each pollutant, so that each square mile could be coded on a scale of from zero to thirty-eight pollution problems. This scale was converted to percentages--that is, the proportions of the thirty-eight pollution problems present in each square mile--and the proportion was then mapped. Areas of high, medium, and low pollution were differentiated.

The sample communities were selected on the basis of this pollution map as well as on the basis of their social rank. The matrix of the study communities is given below.

	High Pollution	Low Pollution
High SES	South Shore Lincoln Park Near North Kenwood Hyde Park	Forest Glen Dunning Montclare Garfield Ridge Clearing Beverly Roseland Calumet Heights
Low SES	Uptown West Garfield Park East Garfield Park North Lawndale Woodlawn Englewood	Burnside Pullman Avalon Park

Study communities were also selected to include both black and white populations in the various status categories as well as to maintain some geographical balance (primarily north and south).

Because the study is concerned with community attitudes, not individual attitudes, it was necessary to ensure that those social characteristics that defined the community were as homogeneous as possible. Since some of the selected communities did not meet this requirement, selected subareas of the study communities were examined (census tracts). Further stratification was needed, and so the highest-status subarea in the high-status community was studied, as was the lowest-status subarea in the low-status community. This resulted in a homogeneous grouping of respondents at the extremes of the socioeconomic scale.

Questionnaire Administration

The sample was drawn by the Institute for Social Action, a professional survey organization, using a reverse telephone directory for the city of Chicago. The number of assigned cases, 1,738, was thought to be adequate to insure attainment of the 1,067 needed to meet the error limits set by the researcher. Some areas were sampled disproportionately because of migration patterns and expectations that more telephone numbers would be disconnected or changed. Table 1 shows the ratio of net to assigned cases for each community. As can be seen, the lowest percentages are found in Woodlawn, because of nonavailability and other reasons such as new residents in the area. This was followed by the communities of East Garfield Park, South Shore, and Englewood, which all had numerous disconnected numbers and language problems (a predominantly Spanish area was selected that fell into the limits of the survey area). Overall, of the 1,738 assigned cases, 1,241 were contacted, for a net percentage of 71.4.

Response rates were calculated from the net number of cases. A list of the response rates by community is also found in table 1. As can be seen, there is some fluctuation in response rates. The overall response rate was 75.7 percent--about average in telephone interviewing. Response rates for the various areas were dependent on a number of factors, including the professionalism of the interviewer and negative reactions to interviewing in general. Also, there were language problems in the sense that some people could not cope intellectually with the questionnaire. There were also a number of problems with respondents who were deaf or who were aged and thus unable to cope with the questionnaire because of senility or other reasons (in one area the sample included a nursing home where the interviews had to be thrown out). And some residents had recently moved into the area and thus could not make any statements about it.

One of the greatest problems was transience. The more

TABLE 1

INTERVIEW RESPONSE RATES BY COMMUNITY

Area	Assigned Cases	Net Cases	Percentage Net Cases/ Assigned Cases	Completions	Breakoffs and Refusals	Percentage Completed
Uptown	256	169	66.0	131	38	77.5
Lincoln Park	66	43	65.2	34	9	79.1
Near North	76	63	82.9	42	21	66.7
Forest Glen	33	23	69.7	20	3	86.9
Dunning	33	29	87.9	26	3	89.6
Montclare	38	32	84.2	22	10	68.7
West Garfield Park	48	34	70.8	24	10	70.6
East Garfield Park	38	22	57.9	18	4	81.8
North Lawndale	33	29	87.9	19	10	65.5
Kenwood	103	72	69.9	64	8	88.9
Hyde Park	81	56	69.1	46	10	82.1
Woodlawn	33	13	39.4	10	3	76.9
South Shore	100	58	58.0	45	13	77.6
Avalon Park	203	159	78.3	115	44	72.3
Burnside	100	56	56.0	31	25	55.3
Calumet Heights	100	87	87.0	70	17	80.5
Roseland	100	81	81.0	67	14	82.7
Pullman	130	91	70.0	60	31	65.9
Garfield Ridge	33	26	78.8	21	5	80.8
Clearing	33	28	84.8	20	8	71.4
Englewood	67	39	58.2	31	8	82.1
Beverly	34	31	91.2	24	7	77.4
Totals or Average	1,738	1,241	71.4	940	301	75.7

transient the area the higher the refusal rate. This was true in Uptown as well as in North Lawndale. The lowest response rates were found in Burnside (a lower-class white ethnic area), North Lawndale (a lower-class black area), Montclare (a middle-class white ethnic area), and the Near North (a wealthy, high-class white area). There was no apparent trend in the rates of refusal.

The number of cases completed was 940. Eight of these were discarded because they fell out of the community areas they were supposed to represent. Five of these were in Uptown, two were in Kenwood, and one was in Pullman.

Since the sample was stratified by a typology, it was necessary to obtain 134 questionnaires per type. In some instances this goal was not reached (table 2). This is particularly important in type 3, the low pollution, low socioeconomic status, white category; hence caution is indicated in analyzing the results of this class. On the community level caution must be exercised with Woodlawn and Burnside, both of which failed to yield even half the required sample cases. As long as these limitations are noted there should be no problem with the validity of the overall analysis.

The survey was administered between 15 May 1976 and 1 July 1976. All the interviewers were female, and the interview time was generally from 5 p.m. to 9 p.m. Monday through Friday and all day and evening (until 9 p.m.) Saturday and Sunday, to insure that some adult would be in the home. The interviewer was told to present the questions to any adult who answered the phone. If a child answered, he was asked to call one of his parents to the phone, to avoid any sex bias. As it happened, more women (66.5 percent) than men (33.5 percent) answered the phone and became respondents; therefore the survey is slightly biased in terms of sex.

TABLE 2

NUMBER OF INTERVIEWS NEEDED BY COMMUNITY

	Number Needed	Actual Number Received	Percentage Received/ Total Needed
Chicago	1,067	932[a]	87.3
Type 1	134	126	97.8
Uptown	134	126	97.8
Type 2	134	102	76.1
West Garfield Park	27	24	88.9
East Garfield Park	27	18	66.7
N. Lawndale	27	19	70.4
Woodlawn	27	10	37.0
Englewood	27	31	100.0
Type 3	134	90	67.2
Burnside	67	31	46.3
Pullman	67	59	88.
Type 4	134	115	85.8
Avalon Park	134	115	85.8
Type 5	134	122	91.0
Lincoln Park	45	34	75.6
Near North	45	42	93.3
Hyde Park	45	46	100.0
Type 6	134	107	79.8
Kenwood	67	62	92.5
South Shore	67	45	67.1
Type 7	134	133	99.2
Forest Glen	23	20	87.0
Dunning	23	26	100.0
Montclare	23	22	95.6
Garfield Ridge	23	21	91.3
Clearing	23	20	87.0
Beverly	23	24	100.0
Type 8	134	137	100.0
Calumet Heights	67	70	100.0
Roseland	67	67	100.0

[a] Adjusted for the eight cases that fell out of their respective community areas.

What do the people of Chicago feel about pollution? This is the topic of part 2. Chapter 5 discusses the results of the questionnaire; chapter 6 considers whether attitudes are related to social characteristics and pollution levels; and chapter 7 centers on the behavioral intent model with the prediction of intent to abate pollution and actual behavior to abate pollution.

PART TWO

CHAPTER V

FREQUENCY AND CONSISTENCY OF
QUESTIONNAIRE RESPONSES

What kind of information was found on people's attitudes about pollution? The first part of this chapter concerns itself with a report on the frequency and types of responses to the survey. The rest of the chapter looks into the relationships between worry rate and seriousness of the pollution problem; seriousness and intent to abate pollution; and worry rate and intent to abate pollution. Characteristics of the entire sample are presented, as well as differences among the eight types of study areas described in Appendix 3: type 1--high pollution, low socioeconomic status, white; type 2--high pollution, low socioeconomic status, black; type 3--low pollution, low socioeconomic status, white; type 4--low pollution, low socioeconomic status, black; type 5--high pollution, high socioeconomic status, white; type 6--high pollution, high socioeconomic status, black; type 7--low pollution, high socioeconomic status, white; type 8--low pollution, high socioeconomic status, black.

Frequency of Responses

Worry about Various Social Problems

We found that crime was the most worrisome problem to the respondents: 62 percent of them worried about it very often. This was not surprising for the city as a whole, but in every type of area examined, crime was one of the three problems most worried about.

TABLE 3

CITYWIDE FREQUENCY OF WORRY ABOUT VARIOUS SOCIAL PROBLEMS (Percentages)

Problem	Very Often	From Time to Time	Not Much	Don't Know
Crime	62.3	23.4	14.1	0.2
Drugs	51.1	21.2	27.1	0.5
Quality of education	50.9	19.8	28.9	0.4
Housing costs	45.8	22.6	31.1	0.4
High taxes	44.2	22.9	32.5	0.4
Pollution	39.7	33.2	26.7	0.4
Neighborhood decay	38.4	21.9	39.2	0.5
Traffic congestion	37.9	23.0	38.8	0.3
Racial problems	32.5	23.2	43.8	0.5
Reliability of transportation	20.2	17.1	62.7	0.1

Table 3 illustrates the worry rate for the entire city by specific social problems. Crime is the leader, followed by drugs and quality of education. Pollution ranked sixth.

In type 1 areas (high pollution low status, white), people worried first about crime (58.7%), followed by decay of the neighborhood (46.8%) and then drugs (46.0%). Pollution ranked fifth in this area type.

Respondents in type 2 areas (high pollution, low status, black) said they worried most about drugs (69.6%), crime (66.7%), and then the quality of education (58.8%). In these areas pollution ranked eighth on the scale.

In Type 3 areas (low pollution, low status, white), crime was ranked first (57.8%), followed by quality of education (43.3%) and drugs (41.1%). Pollution ranked fourth.

Type 4 areas (low pollution, low status, black) also

ranked crime as the problem most worried about (78.3%). This is the highest percentage of agreement on any one category in any of the type. Type 4 area residents felt that quality of education (69.6%) and high taxes (64.3%) were the next most worrisome problems. Pollution ranked sixth.

Type 5 areas (high pollution, high status, white) varied from the pattern established by the others. Here pollution was viewed as the problem people worried about most often (41.0%), followed by the cost of housing (40.2%) and crime (38.5%). There is not as much consensus in the percentages, which shows that there is no prevailing opinion, as was the case in type 4, for example, where 78.3% of the respondents said that they worried most often about crime.

In type 6 areas (high pollution, high status, black) residents worried most about crime (72.9%), followed by housing costs (62.6%) and quality of education (57.9%). Pollution ranked eighth.

Crime was again the leading problem in type 7 areas (low pollution, high status, white), with 60.9 percent worrying about it. This was followed by high taxes (55.6%) and drugs (52.6%). Pollution ranked sixth.

Finally, in type 8 areas (low pollution, high status, black) people worried most about crime (66.4%), followed by quality of education (64.2%) and drugs (59.9%). Pollution ranked seventh.

There appear to be only superficial differences among areas of low and high status with regard to what residents worried about. The low status areas were generally most concerned with crime, as were the high status areas. The only real differences were that in the low status areas there was some concern over the decay of the nieghborhood, while in the high status areas there was worry over pollution and the cost of housing. Otherwise, quality of education, drugs, and crime were the leading problems people worried about.

In a black versus white comparison, we find that blacks worried more about crime, drugs, and quality of education. Whites too worried most about crime, drugs, and education, but they were also concerned with housing cost, pollution, and high taxes.

Seriousness of the Pollution Problem

How do the respondents feel about the seriousness of the pollution problem? More than 40 percent felt it was somewhat serious, whereas 32 percent felt it was not serious at all (table 4). Areas classified as low pollution generally felt it was not a serious problem. For example, in area type 7, more than half the respondents felt pollution was not very serious at all. On the other hand, only one area classified as high pollution felt that it was really a problem. This was area type 2, the low-status black area. The other areas in the high pollution types felt pollution was somewhat serious. Blacks (26.9%) seemed to feel pollution was slightly more serious than did whites (22.6%).

TABLE 4

SERIOUSNESS OF THE POLLUTION PROBLEM
(Percentage of Respondents)

Area	Very Serious	Somewhat Serious	Not Very Serious	Don't Know
Chicago (citywide)	24.2	41.6	32.1	2.0
Type 1: HP, LSES, W	29.4	43.7	27.0	--
Type 2: HP, LSES, B	42.2	39.2	13.7	4.9
Type 3: LP, LSES, W	28.9	53.3	16.7	1.1
Type 4: LP, LSES, B	13.9	42.6	39.1	4.3
Type 5: HP, HSES, W	22.1	40.2	36.9	0.8
Type 6: HP, HSES, B	21.5	43.0	32.7	2.8
Type 7: LP, HSES, W	9.8	34.6	53.4	2.3
Type 8: LP, HSES, B	29.9	40.1	29.2	0.7

Looking at the specific types of pollution most people felt that air pollution was the most serious pollution problem of the four listed (table 5). Also, more than half the respondents felt that water and solid waste pollution were not very serious in their area. In answering question 3, most respondents felt

TABLE 5

SERIOUSNESS OF SPECIFIC POLLUTION PROBLEMS
(Percentage of Respondents)

Type of Pollution	Very Serious	Somewhat Serious	Not Very Serious	Don't Know
Air pollution	32.4	34.0	32.3	1.3
Noise pollution	25.3	27.0	47.0	0.6
Solid waste pollution	22.9	26.0	50.9	0.3
Water pollution	15.7	23.6	55.2	5.6

that their friends and neighbors worried about pollution about as much as they did themselves (63.7%). There was no discernible pattern across area types on this question. Concerning whether pollution bothered them, we found that more than 60% of the respondents were not made angry or depressed by pollution, but 55.7 percent responded that pollution annoyed them. Again, there were no differentiable patterns across area types.

Willingness to Abate Pollution

Exactly what types of actions do the people of Chicago consider taking against pollution? We found that most people (92.6%) were willing to vote for environmental legislation or environmentally concerned candidates, and that this measure was considered somewhat effective in abating pollution. Table 6 provides a ranking of possible actions to abate or avoid pollution. The next most popular action was signing petitions. Again, most people who were willing to do so felt this measure was somewhat effective. These two actions are those most commonly used by citizen/activist/environmental groups in stressing public involvement.

TABLE 6

RANKING OF SPECIFIC ACTIONS TO ABATE OR AVOID POLLUTION

Intent Action	Percentage Willing to Do	Rank
Vote for legislation or candidates	92.6	1
Sign petitions	86.6	2
Avoid swimming in Lake Michigan	71.1	3
Use less electricity	71.0	4
Join a group	70.1	5
Complain to authorities	69.0	6
Attend local gov't meetings	62.8	7
Pay more for goods so industry can recycle	56.8	8
Keep windows closed more	56.1	9
Use a car less	55.2	10
Pay higher taxes	52.8	11
Avoid boating on Lake Michigan	52.7	12
Get out of the city more	46.5	13
Stay indoors more	41.1	14

The third-ranked action was to avoid swimming in Lake Michigan. More than 70 percent of the respondents said they would do this in order to avoid pollution. This measure was also judged the least effective of all the actions listed.

Almost three-fourths of the respondents were willing to reduce their use of electricity, and they again considered this measure somewhat effective. The fifth-ranked action was joining an environmental or neighborhood group concerned with pollution, a measure judged somewhat effective. The measures judged most effective were paying higher prices for goods so that industry could install antipollution equipment and recycle waste materials, and getting out of the city. The measures judged least effective were complaining to authorities and not swimming in Lake Michigan.

Actual Behavior with Regard
to Pollution Abatement

When the respondents were asked whether they had actually carried out any of the intent actions listed in the previous question, 48 percent replied that they had. Looking at the various communities, then the eight area types, we find that more people in the white upper-class areas have done something to abate pollution. The percentages range upward of 60 percent: 87 percent of the Hyde Park respondents have done something to abate pollution. The specific action that the respondents most often mentioned was using less electricity. This was followed by signing petitions and complaining to authorities. Overall, of the 48 percent who mentioned that they took some action, 20 percent did only one thing and 12.2 percent did two.

Among the responses or actions mentioned that were not listed on the questionnaire were the following:

--use unleaded gasoline in car and boat
--serve on commissions
--try to use recycled goods
--contribute financially to environmental groups
--change soaps
--pick people off the street
--not buy colored toilet tissue (more polluting than white)
--buy a garbage compactor
--clean up a prairie in the area
--buy bottles instead of cans
--buy a smaller car
--change from oil to gas heat.

One person who said he had reduced the amount of electricity he used commented, "It doesn't come hard because I learned to make do in the old country." Another respondent replied, "I've discussed it to the point of revolution, but it's all been in conversation. Never have gotten around to doing things."

Reasons for Not Doing More

The primary reasons people gave for not doing more about pollution was that they didn't have enough time (28.3%). This was followed by the responses: "it wouldn't do any good" (19.7%); "I don't know" (16.6%); "too many hassles" (16.0%); "not interested" (7.5%); and combinations of those (11.7%).

There were no distinctions on this question between races, different status classifications, or different pollution levels.

Willingness to Abate Solid Waste Pollution

People are overwhelmingly willing to ask or require people to clean up after their dogs (91.6%). This measure is also considered very effective. A ranking of the specific actions to abate solid waste pollution is found in table 7. The second-ranked action came as a mild surprise: more than 90 percent of the respondents said they would be willing to separate their trash so that cans, bottles, jars, and newspapers could be recycled. This measure was considered somewhat effective to very effective in reducing solid waste pollution.

Voting was considered somewhat effective in abating the problem and ranked third on the solid waste intent list. Picking up litter ranked fourth and was thought to be very effective. And paying more for goods, considered very effective, was ranked fifth, with 81.3 percent of the respondents willing to take this action.

Actual Behavior with Regard to Abating Solid Waste Pollution

When asked whether they themselves had done anything to abate solid waste pollution, two-thirds of the respondents said they had taken some type of action. This is much higher than

TABLE 7

RANKING OF SPECIFIC ACTIONS TO ABATE SOLID WASTE POLLUTION

Intent Action	Percentage Willing to Do	Rank
Ask or require people to clean up after their dogs	91.6	1
Separate trash for recycling	90.8	2
Vote for legislation	89.2	3
Pick up litter on the street	83.9	4
Pay more for longer-lasting goods or products	81.3	5
Complain to authorities	79.1	6
Ask that more trash containers be put on the street	77.1	7
Pay refundable deposit on beverage containers	76.3	8
Stop using disposable products	58.0	9
Pay to have trash separated for recycling	34.1	10

for general pollution. Most of these people had picked up litter on the street; the next-largest group had separated their trash for recycling. It is not clear whether they actually recycled the trash or merely separated the wet or food garbage from the paper waste.

Reasons for Not Doing More to Abate Solid Waste Pollution

There was no clearly distinguishable reason for not doing more about solid waste pollution. Thirty-five percent said this was because there were "too many hassles," but 54 percent said it was not. Forty-four percent replied that time was a factor; another 44 percent said it was not. Thirty-nine percent replied that it would not do any good, but 51 percent said that was not the reason. And 77 percent replied that lack of interest was not a reason for their not doing more.

This response led us to a close examination of question 11: "Are there any other reasons why you have not done more?" We found that 7.3 percent claimed the primary reason was that solid waste was not a problem in their area. Thirty-four percent replied that there were other reasons they had not done more, and 58 percent said there were no other reasons. The range of other answers proved enlightening, and some general trends were apparent in each community.

Other Reasons for Not Doing More
about Solid Waste Pollution

Uptown (Type 1)

Fifty-two percent of the Uptown respondents felt that there were no other reasons they did not do more; 4 percent felt solid waste was not a problem in their area; and the remainder (43.7%) gave other reasons for their lack of action. Among those reasons were:

- --lack of support from public authorities for recycling efforts
- --it must be a collective effort by the residents--one person cannot do it on her own
- --too many sloppy people who do not care enough
- --pollution is not all that important--we should concentrate on other more important matters such as safety and jobs
- --it is the responsibility of people who are supposed to do their jobs
- --people do not care
- --it would be one man against the flood.

Two of the individual responses were:

"It doesn't affect me or anyone in our neighborhood. Our neighborhood is very clean. Two blocks from us is another story, but the reason is the kind of people who live there. Our main worry is crime. We have to be very careful now. This used to be a wonderful neighborhood."

"The suggestions you made aren't going to be very effective. The people in my neighborhood are very poor and the neighborhood is very trashy. The problem is much more complex."

West Garfield Park (Type 2)

West Garfield Park residents replied that the reasons for their lack of action on abating solid waste pollution were that it would do no good (58.3%) and that there was too much red tape (45.8%). More than 79 percent of the respondents felt it was not a problem, and 17 percent felt it was. Responses ranged from "old age and disabled," "I don't have time to lose from work to go downtown to complain," "you need a group and one person cannot do it alone," to cooperation and the assertion that the neighborhood was too dangerous to walk around in. One woman said, "The garbage comes from people driving by and throwing cans of beer or wine out the window."

East Garfield Park (Type 2)

More than 60 percent of the residents claimed that "too many hassles" and its not doing any good were the main reasons for their not doing more; 5 percent felt it was not a problem, and 33 percent gave other reasons that included:

>   --you have to get other people interested
>   --neighbors don't cooperate so I lost interest too
>   --I do participate in picking up garbage, but it's a waste of time because kids are throwing it around faster than you can pick it up.

There was also this response: "Options left to individuals don't work. Petitions don't work. You may get responses and you may not. I have become apathetic to all these problems because officials don't respond and I can't do it alone." The most forceful response in the entire survey was found in this community. The respondent was angry and was more than willing to tell his story to the interviewer:

> I live at city hall, but they are the dead hang of everyone. They don't do nothing but complain to me cause my building has rats and I got rats because other people don't care. There's addicts around here smoking reefers all day and the dirty people under Daley don't do a God damn thing about it.

The police are afraid to get out of their cars. I'm always complaining to city hall, but they don't care. They just want their paychecks. No one does anything about these problems. City is making money on abandoned buildings and won't tear them down so they are full of rats. How can they expect me to keep my building free of roaches and rats when their buildings are left in ruin? I pay for street sweepers but don't get none.

North Lawndale (Type 2)

Seventy-three percent of the residents responded that they had no time to do more about solid waste pollution and that it would not do any good. Some of the other reasons mentioned were health and age, needing more people, and "no one has ever approached me to do these things."

Woodlawn (Type 2)

The four reasons listed accounted for Woodlawn respondents' failure to do more, although most answered "don't know" to all four reasons.

Englewood (Type 2)

The four reasons provided on the questionnaire seemed to cover all the reasons Englewood residents did not do more. Lack of cooperation was also mentioned a few times. One person felt it was up to the precinct captains and ward committeemen. Another person said that if laws were passed requiring people to clean up, then he would be interested.

Burnside (Type 3)

The four reasons provided in the questionnaire did not cover why Burnside residents did not do more. More than 70 percent said there were other reasons. Six percent felt solid waste was not a problem in their area. The other responses ranged from "other people don't care to do anything," "old and disabled," and "need a community effort," to "never thought about it" and "I never thought of pollution in the city." Some other responses were: "It hasn't dawned on a lot of us the apparent seriousness of the problem," and "This neighborhood is 90 percent black and if you complain about the kids throwing bottles and other litter, they mark your gutters with an X and do twice as much damage to you."

Pullman (Type 3)

Eight percent of Pullman residents felt solid waste was not a problem in their area and said this explained their lack of activity in abating it. But 47.5 percent replied that there were other reasons besides the ones listed. Some of these were:

--we've done enough
--we have a civic group that takes care of these things
--I do what I can
--I don't get involved
--it is the responsibility of property owners
--it is not an individual problem, it is an industry problem.

Some of the individual responses were:

"It should be done by the city of Chicago. Whatever happened to street cleaners? They used to clean the streets every day."

"There is no place to recycle things. This should be done on a municipality level. If the city had separate trucks to pick up things, it would be easier to recycle."

Avalon Park (Type 4)

The primary reasons Avalon Park residents gave for not doing more were "lack of time," "no cooperation among neighbors," and "never really thought about it." Seven percent of the respondents felt solid waste was not a problem in their area. One person replied, "I don't see why it's got to be the little guy to do it. Let the city do it. They've got men who stand around doing nothing. Put them to work. Big companies don't want to spend money for cleanup. It's always the little guy who has to run around." And another resident said, "It's not that I haven't done more, but when I look at large companies who are doing the polluting, I feel that they want the working people to pay for it."

Hyde Park (Type 5)

Most Hyde Parkers gave other reasons for not doing more about solid waste pollution. Seventeen percent said it was not a problem and the majority said they had no time to do more. Among the other responses were:

--never thought about it
--I feel it's hopeless in the city
--people are not informed
--no need for it
--I do what I can and what I am supposed to do
--it doesn't offend me
--not enough people are concerned with individual action
--I'm only minding my own business and not stirring up any more than has already been stirred up.

Also, there was, "What's another wrapper? My sloppiness will blend in with my neighbors'. Why start with me?" And "All the approaches are ineffective and not very intelligently carried out--politically oriented rather than community oriented--certain groups that have a specialized interest are competitive with other specialized interests."

Near North (Type 5)

Near North residents primarily said they had no time to do more to abate solid waste pollution. More than 40 percent, however, said their inaction was due to other reasons. These include pollution's not being a problem (7.1%) and the following:

--it's up to the workers for the city
--when I'm going to work I don't feel like picking up litter, but on a weekend I will
--it's more that I don't know what to do, how to go about it, or what it is I'm supposed to do
--I'm not familiar with representatives in my area, the channels to be used to do something
--I have no direction--I don't know where to begin with these problems
--I don't feel at home here--I fell like an outsider and observer
--living in Chicago, I feel it's who you know

--It lies more in the attitude of people rather than
the authorities
--I live in a highrise and I'm new to Chicago. I
don't know too much about these questions
--we do our share.

Two other responses were, "Everything is done for us--I don't have to do anything," and "Garbage is a matter of education--it's the people who make the garbage!"

Lincoln Park (Type 5)

The primary reason given in Lincoln Park was the amount of red tape. Only 8.8 percent felt that pollution was not a problem, and 61.8 percent offered other reasons. These included:

--as an individual I don't feel I have the ability to do
anything
--health reasons
--not aware of the problem
--apathy
--done enough
--the people who are concerned are doing a lot
--there are more important issues
--I'm not a doer but I have a lot of opinions
--It's too dirty--it's germ-ridden, which is why I don't
do more
--I think they have overrated pollution in all phases--I
don't feel there is any need to do more.

Some of the more detailed responses included the following:

"Pollution is a very complex problem. No one has a solution. I feel there are more young people who are becoming involved in it, and that's good."

"I thought about separating my trash, but I never know where to bring it. Besides that, if there is a place for recycling, the hours are too inconvenient for working people."

"I don't think the politicians and big business are interested enough in pollution. They have to make the first move."

"I was unable to find out where to take things for recycling. Perhaps more public information on the subject would be advantageous."

"Why should I do something when no one else cares to do

something, too? Everyone should be required by enforced laws not to litter. They should take disposable products off the shelves. If you don't see them, you won't buy them."

South Shore (Type 6)

The four main reasons given for lack of action in South Shore were lack of time, feeling that it would do no good, "neighbors don't care about the problem," and "no cooperation among the residents." One person said, "We live a secluded life and don't get out often. When we go out, we are scared." And another person replied, "I don't think the public is educated to the importance of all types of pollution problems that affect them or they really don't care."

Kenwood (Type 6)

The dominant response in Kenwood was that the people had no time. Fewer than 4 percent felt it was not a problem, and 34 percent gave other reasons for not doing more:

- age and physical disability
- I do what I can
- not important on my priorities
- it's too political
- if it got bad enough, I'd do something
- lack of understanding of the solutions and problems among people
- no cooperation among neighbors
- politicians overspend and nothing is done about the neighborhoods; have to fight for everything
- you need to know where you can take your papers to be recycled.

One person said, "I don't believe in agencies or bureaucracies! No regulations--pass laws." And one person said, "I think there should be greater concern on the part of people toward a clean environment--by that I mean greater knowledge of the harmful effects of pollution."

Montclare (Type 7)

The primary reasons for not doing more about solid waste pollution given by Montclare residents were "not enough time," "wouldn't do any good," and "too much of a hassle." Nine percent felt that solid waste was not a problem in their area, and 36 percent mentioned other reasons. Among these were:

- it isn't necessary to do more
- physically disabled
- don't go out much
- not concerned with the problem
- people don't care--they have no civic pride
- if others did it, so would I.

And, finally, "You antagonize people if you complain too much."

Dunning (Type 7)

Lack of time seemed to explain why Dunning respondents did not do more about solid waste pollution. Some of the other reasons included "people are too lazy," and "people are not too cooperative--they only do what they have to."

Forest Glen (Type 7)

There was no clear reason given why Forest Glen residents did not do more. Only 20 percent responded that it was because of other reasons, yet none of the reasons provided on the questionnaire were answered affirmatively by a majority. The range of the other answers was:

- teach children not to be abusive would be the best thing
- you have to get complete cooperation to do any good
- we should all do it and there wouldn't be any problem
- it's everyone's problem to take care of pollution.

Garfield Ridge (Type 7)

The reasons given for lack of action on solid waste abatement in Garfield Ridge were: "too many hassles," "it isn't a problem," and "I have already done my share."

Clearing (Type 7)

The primary reason given for not doing more in Clearing was lack of cooperation from neighbors.

Beverly (Type 7)

The leading reason given in Beverly for not doing more was that the residents felt there was no problem. Most of the other responses were: "I do what I can," "you need many people to get anything done," and "I'm not an expert and this should be done by people who know what to do to combat it." Some of the individual responses were:

"I think you have to go to the manufacturers of things such as bottles and disposable products. They have to stop producing things so people won't buy them."

"I really think pollution is overrated in Chicago. It's not as bad as it's made out to be."

Calumet Heights (Type 8)

The main reasons cited for lack of activity in this community were lack of time and "too many hassles." Other important reasons were: "It's not a problem here," "don't know what to do," and "never really thought about it." One person commented, "I am socially aware and so I do take the time to do things that I think would aid and preserve the environment." And another said, "I am following the line of least resistance. If someone started, I would follow."

Roseland (Type 8)

Roseland residents provided their own reasons for not doing more about solid waste pollution. Sixteen percent said it was not a problem. Other reasons included:

>--pollution laws are not enforced, so what's the use
>--others are not interested
>--it should be done by higher authorities
>--I do as much as I can
>--old and disabled
>--need to know the right people
>--in the area I'm in, there are a lot of taverns, and these are the people who won't clean up

Some of the individual responses were:

"People worry too much about everything. We ain't got no more pollution today than we ever had."

"The only way to do away with pollution is to chain people down."

"The Vehicle License Bureau is close to my home and so is Chicago State, and you are losing a battle trying to clean up after the kids from Chicago State."

"I think God takes care of that. Rain and wind takes care of pollution. People are going wild about something they should put in the hands of God."

"Programs are not made known to people. If cans and newspapers would have a deposit on them, people would rush back to the store with them. You don't see them throwing away a bottle when they can get five cents back on it, but why should they save newspapers when it takes so many pounds to get a nickel? People need an incentive."

Willingness to Respond in More Detail

In response to the last question of the interview, a large percentage of those interviewed (76.2%) said they would be willing to respond in more detail on their attitudes toward pollution. The lowest rate was 50 percent, in Dunning. The highest response rate was 100 percent, in three communities: East Garfield Park, Woodlawn, and Burnside. We concluded that people were very interested in this subject and extremely willing to discuss it. This impression was confirmed by the telephone interviewers, who responded that in many instances people would not answer the questions with a simple yes or no but wanted to discuss every answer. This ultimately increased the length of time the interview took. Instead of ten minutes, the interviews averaged between fifteen and eighteen minutes.

Consistency between Worry Rate, Perceived Seriousness, and Intent to Abate Pollution

It was suggested that people who worry very often about pollution would also tend to consider it a serious problem and that people who did not worry much about pollution would consider it less serious. This idea was examined not only for the city as a whole but for each area type, to see whether there were inconsistencies between how often a person worried about pollution and how serious she thought it was and whether this varied by social levels and pollution levels.

It was also suggested that the more serious the respondent felt the problem was, the more willing he would be to do something about it. This was also examined by each type of area for the number of actions (out of fourteen) they said they would be willing to do; and also by actual actions. We hoped that some pattern of perceived seriousness would emerge in relation to a specific action by a specific group (differentiated by pollution level, race, and status).

Finally, we compared worry rate and the number of intended actions, to see whether worry influenced the number and kinds of actions the respondent would take to abate pollution.

The results of these cross-tabulations are presented in tables 8-16. The percentages falling in each cell are given for the citywide situation only. Citywide chi-square and gamma statistics are provided, and variations from the citywide patterns are discussed in the text. Finally, a first-order partial gamma was computed while controlling for pollution level, status, and race for general pollution and for the specific types of pollution--air, water, and solid waste. If the first-order partial gamma approaches 0.00, then the relationship considered in the citywide case can be explained by social and environmental differences. If it increases from the zero-order gamma statistic (approaches 1.00), the relationship can be considered citywide and not socially or environmentally specific.

Worry Rate and Perceived Seriousness of Pollution

In the examination of worry versus perceived seriousness of the problem (table 8), the contingency table suggests a weak relation among concordant pairs. (gamma = .37, p = .00). In other words, people who think about pollution frequently also consider it a serious problem, whereas people who do not worry about pollution very often do not consider it a very serious problem.

TABLE 8

WORRY AND PERCEIVED SERIOUSNESS OF POLLUTION

Frequency of Worry	Percentage Very Serious	Percentage Somewhat Serious	Percentage Not Very Serious	Don't Know
Very often	14.1	16.8	8.4	0.4
From time to time	5.2	16.0	10.5	0.4
Not very often	4.0	8.7	13.0	1.1
Don't know/ no response	--	0.1	0.2	0.1
	23.3	41.6	32.1	2.0

NOTE: Significance = .000; chi-square = 94.49; gamma = .37.

When we control for pollution levels we again find little variation from the citywide pattern (table 9) in terms of the gamma statistic, though the chi-square statistic has decreased somewhat. In controlling for the race of the respondents very little variation in the statistics is evident. But, in the status variable control we do find some distinctions between the high-status group and the low-status group. The high-status group has a lower chi-square value, indicating a less systematic relationship between worry rate and perceived seriousness, and also a slightly lower gamma statistic, indicating basically no relationship between the concordant cells in the contingency table. The low-status group shows a higher degree of association between worry rate and perceived seriousness.

TABLE 9

SUMMARY STATISTICS: FREQUENCY OF WORRY AND PERCEIVED SERIOUSNESS OF POLLUTION

Citywide	Gamma .37	Chi-Square 94.50	Significance .000
Control for Pollution			
High pollution	.35	55.91	0
Low pollution	.42	52.52	0
First-order gamma	.39	--	--
Control for Status			
High status	.24	34.12	.0001
Low status	.49	75.40	0
First-order gamma	.39	--	--
Control for race			
White	.40	50.86	0
Black	.36	47.83	0
First-order gamma	.38	--	--

The first-order partial gamma statistic increases from the zero-order statistic in all three control situations; so we can conclude that the relationship between worry and perceived seriousness of pollution is not influenced by pollution level, status, or race.

Worry rate and perceived seriousness
of air pollution

We decided to look at each individual pollutant to see whether this trend held as it did for general pollution. In the case of air pollution (table 10) there were strong indications that people either worried very often about air pollution and considered it a serious problem or worried very little and considered it not very serious. The gamma statistic here is higher (.48) than in the general pollution case. There is an even split over seriousness between those who consider air pollution very serious (32.4%) and those who feel it is not very serious (32.3%), as well as an even split in the worry category.

TABLE 10

WORRY AND PERCEIVED SERIOUSNESS OF AIR POLLUTION

Frequency of Worry	Percentage Very Serious	Percentage Somewhat Serious	Percentage Not Very Serious	Don't Know
Very often	19.8	12.0	7.4	0.4
From time to time	8.6	15.1	9.2	0.2
Not very often	3.9	6.9	15.5	0.5
Don't know/ no response	0.1	--	0.2	0.1
	32.4	34.0	32.3	1.2

NOTE: Significance = .000; chi-square = 177.90; gamma = .48.

Areas 2 (HP, LSES, B) and 4 (LP, LSES, B) reverse the trend found in the citywide air pollution case; that is, people who do not worry about pollution and feel it is not a serious problem predominate (14.7% in area 2 and 20.0% in area 4). This is followed by "worry not very often," "somewhat serious" and "worry often," "very serious" in area 2 and "worry often," "not serious" and "worry often," "somewhat serious" in area 4. The only factor common to both these types is that they are both black and low status. This may account for the lack of worry about pollution and the evaluation of it as not serious, since other pressing social problems may have a higher priority for the population.

As was found for general pollution, the first-order partial gamma statistic (table 11) either stayed the same as the zero-order statistic or increased slightly in this specific case. This leads to the conclusion that race, status, and actual pollution level did not affect the basic relationship between worry rate and perceived seriousness of air pollution.

TABLE 11

SUMMARY GAMMA STATISTICS: FREQUENCY OF WORRY AND PERCEIVED SERIOUSNESS OF SPECIFIC POLLUTION

	General Pollution	Air	Water	Noise	Solid Waste
Citywide	.37	.48	.28	.22	.15
Control for Pollution					
High pollution	.35	.50	.31	.24	.20
Low pollution	.42	.49	.26	.23	.20
First-order partial gamma	.39	.49	.28	.23	.18
Control for Status					
High status	.24	.44	.25	.17	.09
Low Status	.49	.52	.30	.26	.25
First-order partial gamma	.39	.48	.28	.22	.17
Control for Race					
White	.40	.48	.23	.22	.08
Balck	.36	.48	.33	.21	.23
First-order partial gamma	.38	.48	.28	.22	.16

Worry rate and perceived seriousness of water pollution

In the case of water pollution one cannot discern any trend other than to say that most Chicago residents feel that water pollution is not a very serious problem (table 12). This observation holds up for the individual area types; that is, water pollution is not judged serious and people worry at different rates about the problem, with no statistical trend apparent. Also the worry rate and perceived seriousness of the problem are not specific to environment, status, or race (table 11).

TABLE 12

WORRY AND PERCEIVED SERIOUSNESS OF WATER POLLUTION

Frequency of Worry	Percentage Very Serious	Percentage Somewhat Serious	Percentage Not Very Serious	Don't Know
Very often	8.9	11.2	16.8	2.8
From time to time	4.8	8.	18.2	1.8
Not very much	1.9	4.2	19.7	0.9
Don't know/ no response	--	--	0.3	0.1
	15.6	23.7	55.0	5.6

NOTE: Significance = .000; chi-square = 67.60; gamma = .28.

Worry rate and perceived seriousness of noise pollution

Basically there are very few trends except that the bulk of the population feels that noise pollution is not a serious problem (table 13). This is apparent regardless of how often one worries about the problem of pollution. Again, it seems that the rate of worry and perceived seriousness of the problem are not specific to race, status, or environment (table 11).

Most of the areas followed this pattern when each was examined individually. However, area 2 (HP, LSES, B) shows no statistical significance, and the table on worry and noise pollution shows a random ordering with a gamma = .07. Area 1

(HP, LSES, W), on the other hand, shows mild trending (gamma = .31), with people worrying from time to time about pollution and finding noise pollution somewhat serious (19%), worrying very often and finding noise pollution very serious (17.5%), and not worrying much about pollution and finding it not a serious problem (11.9%).

TABLE 13

WORRY AND PERCEIVED SERIOUSNESS OF NOISE POLLUTION

Frequency of Worry	Percentage Very Serious	Percentage Somewhat Serious	Percentage Not Very Serious	Don't Know
Very often	13.6	8.9	16.9	0.2
From time to time	7.0	12.0	14.0	0.2
Not very much	4.6	6.1	15.9	0.1
Don't know/ no response	0.1	--	0.3	--
	25.3	27.0	47.1	0.5

NOTE: Significance = .000; chi-square = 47.99; gamma = .22.

<u>Worry rate and perceived seriousness of solid waste pollution</u>

With solid waste pollution there is no apparent trend in the basic relationship between worry rate and perceived seriousness of the problem. The population feels that solid waste is not a serious problem regardless of how often they worry about it (table 14).

In examining the eight area types, this trend is manifested as discordant pairs of the variables dominant in the contingency table. Also, we find that as in the previous types of pollution, status, race, and pollution level do not tend to alter this basic relationship (table 11).

TABLE 14

WORRY AND PERCEIVED SERIOUSNESS OF SOLID WASTE POLLUTION

Frequency of Worry	Percentage Very Serious	Percentage Somewhat Serious	Percentage Not Very Serious	Don't Know
Very often	11.9	9.7	18.0	0.1
From time to time	5.5	10.0	17.6	0.1
Not very often	5.4	6.3	14.9	0.1
Don't know/ no response	0.1	--	0.3	--
	22.9	26.0	50.8	0.3

NOTE: Significance = .006; chi-square = 22.73; gamma = .15.

Thus it appears that our hypothesis that people who worry often about pollution would also consider it a serious problem is conditionally confirmed--it is somewhat true for pollution generally, and it is true for air pollution. On the other hand, for noise, water, and solid waste pollution, the overriding influence on this relationship is the lack of seriousness of the problem, regardless of how often a respondent worries about the it. The basic relationship between concordant pairs is not influenced by race or by pollution levels. Status, however, does tend to alter this basic relationship for overall pollution and to a lesser degree with solid waste and noise pollution.

Perceived Seriousness of Pollution and Intent to Abate Pollution

In examining these two relationships we were interested in whether a person who felt pollution was serious would also be willing to do more to abate it than one who felt it was not a serious problem. As can be seen from table 15, most respondents fell into the medium-intent category.

TABLE 15

PERCEIVED SERIOUSNESS OF POLLUTION AND INTENT TO ABATE POLLUTION
(Percentage in Each Category[a])

Seriousness	Number of Intent Actions			
	10-14 (High)	5-9 (Medium)	1-4 (Low)	0 (None)
Very serious	11.9	10.5	1.8	0
Somewhat serious	14.5	21.2	5.4	0.5
Not very serious	6.2	17.9	7.3	0.6
Don't know/ no response	0.4	0.5	1.0	0.1
	33.0	50.1	15.5	1.2

NOTE: Significance = .000; chi-square = 82.02; gamma = .38.

[a]Percentages do not sum to 100 due to rounding errors.

There appears to be no significant trending with regard to perceived seriousness and number of intent actions. We find that 21 percent of the population consider pollution somewhat serious and are in only the medium-intent level--a consistent finding. Also, 18 percent of the respondents felt pollution was not very serious yet also fell into the medium-intent category. Based on our original hypothesis, this might be considered inconsistent. Finally, 14.5 percent of the respondents felt pollution was somewhat serious yet fell into the high-intent category. Here again we would consider this finding inconsistent. Thus it appears that there is no relationship between how serious one finds pollution and the number of intent actions one would carry out to abate the problem.

In the control situation, we find that actual pollution level, status, and race do not significantly influence this basic relationship (table 16).

TABLE 16

SUMMARY STATISTICS: PERCEIVED SERIOUSNESS OF POLLUTION
AND INTENT TO ABATE POLLUTION

Citywide	Gamma .38	Chi-square 82.02	Significance .000
Control for Pollution			
High pollution	.42	52.80	0
Low pollution	.34	46.88	0
First-order gamma	.38	--	--
Control for Status			
High status	.38	41.81	0
Low status	.39	60.00	0
First-order gamma	.39	--	--
Control for Race			
White	.38	38.04	0
Black	.37	54.10	0
First-order gamma	.38	--	--

Worry Rate and Intent to Abate Pollution

It was suggested that a person who worried very often about pollution would be willing to perform a large number of intent actions to abate the problem and that a person who worried very little about it would be willing to carry out few or no intent actions to abate pollution.

We found that this was somewhat true. The largest percentage cell in the contingency table was composed of people who worried often about pollution and who said they would be willing to carry out five to nine of the fourteen intent actions listed (table 17). The next highest percentage included those who worried often about pollution and were willing to take ten to fourteen of the intent actions and those who worried from time to time about pollution and were willing to take five to nine of the intent actions. As can be seen, these last two are indeed consistent with what we expected. This comparison is significant (p = .000), and the gamma statistic shows a mild trending among concordant pairs in the table.

TABLE 17

WORRY ABOUT POLLUTION AND INTENT TO ABATE POLLUTION

Frequency of Worry	Number of Intent Actions			
	10-14 (High)	5-9 (Medium)	1-4 (Low)	0 (None)
Very often	17.1	19.5	3.1	0.0
From time to time	10.9	17.3	4.6	0.3
Not very often	5.0	13.2	7.5	1.0
Don't know/ no response	0.0	0.2	0.2	0.0
	33.0	50.2	15.4	1.3

NOTE: Significance = .000; chi-square = 86.62; gamma = .38. Percentages do not sum to 100 due to rounding errors.

In a black/white, low status/high status, and low pollution/ high pollution comparison (table 18) there are no distinctions, and the trends appear essentially the same as for citywide general pollution.

Perceived Seriousness of Pollution and Specific Actions to Abate Pollution

In comparing the seriousness of the pollution problem with specific actions a person might take (table 19), we find the following results: a person is most likely to vote and sign petitions as a measure of pollution abatement regardless of how serious she feels the problem is. In all three of the seriousness categories, voting ranked first, followed by signing petitions--individual or passive measures. The third-ranked action for the very serious category was joining an environmental

TABLE 18

SUMMARY STATISTICS: RATE OF WORRY AND
INTENT TO ABATE POLLUTION

	Gamma	Chi-Square	Significance
Citywide	.38	86.62	.000
Control for Pollution			
High pollution	.43	52.83	0
Low pollution	.33	42.62	0
First-order partial gamma	.38	--	--
Control for status			
High status	.35	38.68	0
Low status	.41	55.10	0
First-order partial gamma	.38	--	--
Control for race			
White	.41	51.49	0
Black	.38	49.10	0
First-order partial gamma	.39	--	--

or neighborhood group, whereas for the other two categories it was using less electricity. Using less electricity was ranked fourth by the very serious group. In the somewhat serious category, not swimming in Lake Michigan was the fourth-ranked action, while in the not very serious group, complaining ranked fourth. Complaining was the fifth-ranked action in the very serious and somewhat serious groups, and not swimming in the lake ranked fifth in the not very serious group.

What happens to the rankings when subsets of the population are examined? In the high- and low-pollution comparisons (table 20), we find that voting is still ranked first, followed by signing petitions, not swimming in Lake Michigan, using less electricity, and, finally, complaining. In the low-pollution

TABLE 19

RANKING OF SPECIFIC ACTIONS TO ABATE OR AVOID POLLUTION
BY PERCEIVED SERIOUSNESS OF THE PROBLEM

Intent Action	Very Serious		Somewhat Serious		Not Very Serious	
	%	Rank	%	Rank	%	Rank
Complain to authorities	17.6	5	27.0	5	18.0	
Stay indoors more	11.7	14	17.8	14	11.8	10, 9
Avoid swimming in Lake Michigan	17.3	6	27.3	4	17.3	5
Sign petitions	21.4	2	34.7	2	22.4	2
Use a car less	13.6	9	20.1	10	11.8	9, 10
Pay more for goods so industry can recycle	13.0	10	19.5	11	10.9	11, 12
Join a group	18.5	3	26.7	6	14.7	6
Avoid boating on Lake Michigan	12.4	12	17.1	13	10.8	13
Keep windows closed more	16.1	7, 8	22.7	8	13.9	8
Attend local gov't meetings	16.1	8, 7	24.4	7	14.2	7
Get out of city more	12.7	11	18.5	12	10.4	14
Pay higher taxes	12.3	13	21.4	9	10.9	12, 11
Use less electricity	18.3	4	28.2	3	20.1	3
Vote for legislation or candidates	22.1	1	37.2	1	25.2	1

case the ranking is voting, signing petitions, using less electricity, complaining, and joining a group. The only major distinction between these two groups was the willingness on the part of the low pollution group to keep their windows closed (57.1%) to avoid pollution, whereas only 49.7 percent of the high pollution group said they would use this measure.

In the status comparison we find very little disagreement in the ranking of actions to abate or avoid pollution, with voting, signing petitions, complaining, using less electricity, and not swimming in Lake Michigan the first five actions chosen.

Finally, in our race comparisons, whites ranked voting, signing petitions, using less electricity, complaining, and not swimming in Lake Michigan as the first five ranked actions they would be willing to take. Blacks ranked voting, signing petitions, joining a group, not swimming in Lake Michigan, and complaining as their top five actions, but when we look at some of the other actions, we find some difference in the affirmative responses. More than half the black respondents said they would avoid boating on Lake Michigan, while only 31% of the white respondents said they would use this measure. Sixty-one percent of the black respondents replied that they would attend local government meetings, while only 50% of the white respondents chose this as a measure to abate pollution. Finally, blacks have more of a tendency to join groups to cope with the problem (69%) than do whites (52%).

Thus, for the entire city as well as for specific subsets of the population, voting was regarded as the main measure people would use to abate pollution. This was followed by signing petitions. This ranking did not vary by pollution level, status of the respondents, or race, but we did find some interesting variations by these social and environmental categories in the lower-ranked actions people are willing to take to abate

TABLE 20

RANKING OF SPECIFIC ACTIONS TO ABATE OR AVOID POLLUTION, BY CONTROL CATEGORIES

Actions	Pollution				Status				Race			
	High		Low		High		Low		White		Black	
	%	Rank	%	Rank	%	Rank	%	Rank	%	Rank	%	Rank
Complain to authorities	60.6	5	66.7	4	67.3	3	59.6	5.6	60.9	4	66.6	5
Stay indoors more	35.9	14	41.7	13	39.1	14	38.6	13	32.3	13	45.6	12
Avoid swimming in Lake Michigan	65.2	3	60.6	6	62.9	5	62.8	4	58.0	5	67.9	4
Sign petitions	80.1	2	78.7	2	79.4	2	79.4	2	76.4	2	82.4	2
Use a car less	49.0	10	42.7	9	47.1	12	44.3	10	45.2	11	46.4	11
Pay more for goods so industry can recycle	44.0	11	44.2	10	47.5	10	40.2	12	46.1	10	42.1	14
Join a group	59.5	6	61.7	5	61.5	6	59.6	6.5	52.2	6	69.2	3
Avoid boating on Lake Michigan	38.9	13	42.9	11	38.7	13	43.6	11	31.6	14	50.5	9
Keep windows closed more	49.7	8	57.1	7	53.3	8	53.6	8	49.0	8	57.9	8
Attend local gov't meetings	58.9	7	52.0	8	56.7	7	53.8	7	49.9	7	61.0	7
Get out of the city more	42.0	12	41.9	12	45.3	11	38.1	14	36.9	12	47.1	10
Pay higher taxes	49.5	9	41.3	14	44.7	9	46.0	9	46.3	9	44.3	13
Use less electricity	65.0	4	70.1	3	67.1	4	68.1	3	68.8	3	66.4	6
Vote for legislation or candidates	89.5	1	82.1	1	85.4	1	86.1	1	86.2	1	85.2	1
	N = 457		N = 475		N = 499		N = 433		N = 471		N = 461	

pollution.

Now that we have a picture of the questionnaire responses and the consistency among those responses, what about the main thesis of this research? What are the relationships between attitudes and social characteristics of the community, between attitudes and environmental quality? These questions are taken up in the next chapter.

CHAPTER VI

ATTITUDES, SOCIAL DIFFERENTIATION, AND
ENVIRONMENTAL QUALITY

As we saw in chapter 2, there is a dearth of information on the subject of attitudes, social differentiation, and environmental quality. Awareness of and concern about pollution have been related to exposure levels (ambient environmental quality), but there have been no consistent findings regarding concern about pollution and social indicators or whether social characteristics influence a person's willingness to alleviate the problem.

One of the major flaws of the previous research has been the confusion over the terms <u>attitudes</u>, <u>concern</u>, <u>perception</u>, and <u>opinion</u>. As has already been explained, this study is concerned with attitudes in the strict psychological sense. The computation of attitudes toward pollution and pollution abatement is soundly based in theory (Fishbein and Ajzen 1975).

In correlating attitudes with social differentiation and environmental quality, one must first determine the attitudes from the questionnaire information. This was done in accordance with the theory, as discussed in chapter 3. Attitudes toward pollution in general, air pollution, water pollution, noise pollution, and solid waste pollution were calculated using Fishbein's model. Also, attitudes toward general pollution abatement, air pollution abatement, water pollution abatement, noise pollution abatement, and solid waste pollution abatement also were calculated.

What exactly are the factors that influence a person's attitude toward pollution? Are they mainly social or mainly environmental, or is there some other explanation? Before answering these questions, let us look at a brief description of attitudes in each community and their meaning.

The mean attitude for the entire community area was used, to reduce the difficulties with the data. Since we were interested in community attitudes and our social and environmental information was by community, this posed no problem. Table 21 shows the mean score for each community on attitudes toward pollution abatement. Scores on attitude toward pollution were divided into three major categories: low concern, medium concern, and high concern about pollution. Low or no concern scores (-9 to -2) were defined as communities where the residents did not worry about pollution, did not consider it a problem, and were not bothered by it. High concern scores (+4 and greater) were defined as communities that considered pollution a serious problem, were bothered by it, and frequently worried about it. Medium or indifferent concern scores (-1 to +3) were defined as communities between these two extremes.

On attitude toward pollution abatement, scores of +7 to +19.99 were considered negative attitudes--that is, residents were not willing to do much to abate pollution and found their actions ineffective. Scores of +20 to +32.9 were considered neutral attitudes--residents were somewhat willing to abate pollution and found these measures somewhat effective. And scores greater than +33 were considered positive attitudes-- residents were willing to take action to alleviate the problem and also considered these measures effective.

Attitudes toward Pollution and Social Differentiation

The first task was to ascertain whether there was any association between attitudes toward pollution and the various social characteristics of the communities: a simple contingency

TABLE 21

MEAN ATTITUDES TOWARD POLLUTION AND POLLUTION ABATEMENT

	Community	Pollution	Category	Pollution Abatement	Category
Type 1	Uptown	12.85	High concern	17.75	Negative
Type 2	Englewood	10.10	High concern	36.45	Positive
	Woodlawn	6.90	High concern	15.80	Negative
	North Lawndale	9.42	High concern	39.37	Positive
	East Garfield Park	14.83	High concern	37.79	Positive
	West Garfield Park	9.54	High concern	7.67	Negative
Type 3	Pullman	3.73	Indifferent	14.10	Negative
	Burnside	13.84	High concern	32.52	Neutral
Type 4	Avalon Park	0.43	Indifferent	21.50	Neutral
Type 5	Hyde Park	8.98	High concern	18.98	Negative
	Near North	6.43	High concern	15.76	Negative
	Lincoln Park	6.79	High concern	19.00	Negative
Type 6	Kenwood	4.60	High concern	25.28	Neutral
	South Shore	6.80	High concern	29.91	Neutral
Type 7	Beverly	-7.67	No concern	22.75	Neutral
	Forest Glen	-9.75	No concern	14.90	Negative
	Montclare	-1.50	No concern	14.82	Negative
	Dunning	-4.04	No concern	15.38	Negative
	Garfield Ridge	2.86	Indifferent	21.71	Neutral
	Clearing	0.60	Indifferent	9.70	Negative
Type 8	Roseland	13.70	High concern	22.21	Neutral
	Calumet Heights	1.87	Indifferent	44.08	Positive

analysis was performed to determine this. The variables used in the social analysis are listed in table 22.

Four variables were found to have significant associations with attitudes toward pollution: percentage of single-unit, detached houses, white ethnicity, mean value of houses, percentage of the population owning their homes.

Communities with a small percentage of single-unit, detached houses were found to have a high level of concern about pollution (gamma=.64, p=.01). A high concern score was associated with a negative white ethnic score. In other words, the smaller the white ethnic community, the higher the level of concern about pollution (gamma=.87, p=.02). We found that 45 percent of the communities with fewer than 40 percent white ethnics had high levels of concern.

The association between the mean value of the houses and a community's attitude showed no apparent trending (gamma=.33, p=.002). The other housing characteristic, percentage of the population owning their homes, showed a significant association (gamma=.74, p=.04). It was found that the lower the percentage of homeowners in the community, the higher the concern about pollution.

What can be concluded from the foregoing? It appears from the contingency analysis that people living in communities with a large percentage of apartments, with very few people owning their homes, with homes valued at less than $20,000, and with few white ethnics, had positive attitudes about pollution. That is, pollution bothers them, they feel it is a serious problem, and they frequently worry about it. Does this generalization hold true when the association is examined in terms of direction and magnitude?

Spearman's rank correlation was used to test whether the generalizations in the contingency analysis held true. Spearman's

TABLE 22

SOCIAL VARIABLES

Computer Code	Meaning
SEXRATIO	Sex ratio (male to female)
DENSITY	Density (hundred/square mile)
MFAMINC	Mean family income (dollars)
FMUH	Financing of multiple-unit housing (conventional versus federal financing)
FSUH	Financing of single-unit housing (conventional versus federal financing)
SECOCC	Percentage engaged in secondary occupations
QQCC	Percentage engaged in quinary or quaternary occupations
MOBIL	Mobility (percentage of families living in the area in 1970 who lived in a different house in 1965)
AGE	Stage in life cycle
SUDH	Percentage of single-unit, detached houses
PUTRAN	Percentage of labor force taking bus, subway, el, or railroad to work
PRTTRAN	Percentage of labor force using private transportation to work)
HOUSEAGE	Age of housing
POVERTY	Poverty index (consists of males 16-21, unemployed, not high-school graduates; total income from public payments; units without available automobile; units with more than 1.01 person per room; units without telephone)
BLACK	Percentage black
SPAN	Percentage Spanish
WETH	Percentage of elementary school children in parochial schools (a surrogate measure for white ethnicity)
COMMUTE	Commuters to Chicago central business district
POPCHNG	Population change ratio, 1970/1960
SCHOOL	Median school years completed

TABLE 22--Continued

Computer Code	Meaning
HOMEVAL	Mean value of houses (dollars)
RENT	Mean rent per month (dollars)
OWNHOME	Percentage owning their homes
FIRES	Number of fires per year
LEADPOI	Number of reported lead poisonings per year
MURDERS	Number of murders per year
DEMOS	Number of houses demolished per year
RATBITE	Number of rat bites per year
SOCINDX	Social index (sum of FIRES, LEADPOI, MURDERS, DEMOS, RATBITE)

rho (r_s) can range from -1 to +1, and these extremes indicate perfect matching of the ranks. The closer to zero the correlation, the less obvious the systematic ranking, and therefore at zero there is no ordering or ranking pattern. Rank correlation is a useful measure when variables are in categoric form with some hierarchy and when one wishes to perform more than the perfunctory cross-tabulation analysis.

Nineteen out of the twenty-nine social variables were found to be significant ($p<.05$) in relating social characteristics to attitudes. Those that were not significant were: sex ratio, financing of multiple-unit housing, percentage with secondary occupation, age of housing, percentage using private transportation, and commuters to the central business district. An examination of the correlation vector (table 23) illustrates the same general pattern found in the contingency analysis. Positive attitudes (e.g., high concern) are negatively associated with family income, single-unit detached housing, white ethnic areas, white-collar occupations, rent, percentage owning their homes, and age. The age variable was subsequently dropped because of the way it was measured (four categories representing percentage under eighteen, percentage over sixty-four, both of these, or none of these).

As can be seen, the higher the concern about pollution, the lower the percentage of residents in quinary or quaternary (white-collar) occupations ($r_s=-.52$), the fewer people owning homes ($r_s=.67$) and finally, the smaller the percentage of children attending parochial schools ($r_s=.62$). This indicates that apartment-dwellers are more concerned with pollution than are people who live in single-family houses. Blacks are more concerned than whites. Poverty areas are more concerned about pollution than non-poverty areas, and socially poor areas (high social index) are more concerned and find the problem more serious than socially

TABLE 23

SPEARMAN RANK-ORDER CORRELATION MATRIX: ATTITUDE
TOWARD POLLUTION AND SOCIAL CHARACTERISTICS

Social Variable	Correlation Coefficient (r_s)
Density	.52**
Mean family income	-.43*
Percentage in quaternary and quinary occupations	-.52**
Mobility	.50**
Stage in life cycle	-.50**
Financing of single-unit housing	.42*
Percentage of single-unit, detached housing	-.55*
Poverty index	.43*
Percentage black	.40*
White ethnicity	-.62**
Percentage of labor force using public transportation	.51**
Mean rent	-.36*
Percentage owning their homes	-.67**
Fires	.54**
Lead poisonings	.52**
Murders	.54**
Residences demolished	.52**
Rat bites	.52**
Social index	.54**

*$p < .05$.

**$p < .01$.

rich areas (low social index).

When we control for pollution level, we find that white ethnicity dictates pollution attitudes in an inverse direction. The higher the percentage of white ethnics, the less concerned the community is about pollution (r=-.46, p=.02). Also, as the value of the home decreases, one's attitude toward pollution becomes more positive (r=-.38, p=.04). We also find that the more federal financing of single-unit housing (r=.36, p=.05) and multiple-unit housing (r=-.38, p=.04), the more concern about pollution.

Thus we find for the city of Chicago these results that run contrary to the literature:
1. high concern (positive attitude) negatively correlated white-collar occupations;
2. high concern negatively correlated with white ethnic areas;
3. high concern negatively correlated with percentage owning homes;
4. while controlling for pollution level, high concern negatively correlated with white ethnicity;
5. while controlling for pollution level, high concern negatively correlated with value of homes.

Note that we are able to differentiate attitudes toward pollution on the basis of white ethnicity and the value of the homes regardless of the pollution level.

Attitudes toward Pollution Abatement and Social Differentiation

In the contingency table analysis, three social variables were significant and indicated some relationship with attitude toward pollution abatement. These were percentage black, lead poisonings, and rat bites. Percentage black has a very high association with attitudes toward pollution abatement (gamma=.63, p=.03). Thirty percent of communities with a population less than 10 percent black have a negative attitude, whereas areas with more than 50 percent black have a positive attitude about pollution abatement. Lead poisonings and rat bites also showed some association with attitudes. The relationship was more apparent with lead poisonings (gamma=.60, p=.013), where no cases of lead poisoning were associated with negative or neutral attitudes toward pollution abatement. This was also true with rat bites, although the gamma statistic was smaller (.39), and the significance level lower (.05).

When we use the Spearman rank-order coefficient, these relationships become clearer. The most significant determinant of attitude toward pollution abatement is race (table 24). The higher the percentage black, the more positive the attitude. Occupation is also important. Secondary occupation (blue-collar) is negatively related to attitude, indicating negative attitudes toward pollution abatement by blue-collar workers.

Use of public transportation was found to be positively correlated with willingness to do something about pollution, as was male-to-female sex ratio. The higher the proportion of males in a community, the more positive the attitude about doing something about pollution. Finally, the remaining four variables dealt with the social situation in the community. The more socially depressed areas have a more positive attitude toward pollution abatement and are willing to do more to combat it.

When we control for pollution level, we find very little change in the pattern of social distinctions (table 24). However, in the control situation, density attains statistical significance, but occupation no longer does. Residents of low-density areas are more willing to abate pollution than are those who live in high-density areas (r=-.46, p=.02). In the control situation we also find another aspect of the social environment of the area--rat bites--becoming important and positively correlated with abatement attitudes.

Generalizing from this information, we can make a number of observations. White-collar workers tend to have a positive attitude, while blue-collar workers tend toward the negative.

TABLE 24

SPEARMAN RANK-ORDER CORRELATION MATRIX: ATTITUDE TOWARD POLLUTION ABATEMENT AND SOCIAL CHARACTERISTICS

Social Variable	Correlation Coefficient (r_s)
Not controlling for pollution levels	
Sex ratio	.40*
Percentage in secondary occupations	-.41*
Financing of single-unit housing	.41*
Percentage black	.48**
Percentage using public transportation	.36*
Fires	.41*
Lead poisonings	.48**
Residences demolished	.41*
Social index	.46*
Controlling for pollution level	
Sex ratio (male to female)	.40*
Density	-.46*
Financing of single-unit housing	.44*
Percentage black	.47*
Fires	.50**
Lead poisonings	.52**
Rat bites	.43*
Social index	.48*

Blacks are more willing to take action to abate pollution than whites. And residents of socially depressed or poor areas (defined by the social index) are more apt to have positive attitudes than people living in socially rich areas. Furthermore, as an overall generalization, we find two different subsets of the population that exhibit positive attitudes toward pollution abatement and willingness to participate: the white-collar black population and the socially depressed black areas.

The implications of these findings in light of the current literature on the subject will be discussed in chapter 8. What about the second part of the thesis? What kinds of environmental variables influence attitudes toward pollution (AO) and attitudes toward pollution abatement (AB)?

Attitudes toward Pollution and Environmental Quality

The same procedures were used in the analysis of environmental quality and attitudes as were used for the social differentiation variables. Originally sixty-five variables were used in the analysis, but this was trimmed to sixty because some of them had only one category (i.e., they met the standard or they did not meet it. The variables used in the environmental quality analysis are listed in table 25.

We decided that the logical starting point would be an examination of general pollution. But we will also look at specific pollutants and attitudes toward them.

Seven environmental quality variables were found to have a significant relationship to attitudes toward pollution. From the contingency table, we found that there were high concern scores (positive attitudes) in those areas that met the CO standard for 1974 (in other words, there was a discordant pair relationship that resulted in gamma=-.62, p=.006). We also found that the air quality index was related to concern (statistically

TABLE 25

ENVIRONMENTAL QUALITY VARIABLES

Computer Code	Meaning
SO74	Sulfur dioxide, 1974 (meet or exceed federal standards
SO75	Sulfur dioxide, 1975 (meet or exceed federal standards
PART74	Particulates, 1974 (meet or exceed federal standards
PART75	Particulates, 1975 (meet or exceed federal standards
CO74	Carbon monoxide, 1974 (meet or exceed federal standards
CO75	Carbon monoxide, 1975 (meet or exceed federal standards
NO74	Nitrogen oxides, 1974 (meet or exceed federal standards
NO75	Nitrogen oxides, 1975 (meet or exceed federal standards
OZONE	Ozone, 1975 (meet or exceed federal standards
AIRINDX	Air index cumulative (number of violations of the standards for each parameter)
INDX 75	Air index for parameters measured in 1975
HG	Mercury in the atmosphere (nanograms/m^3), 1971
AS	Arsenic in the air (nanograms/m^3), 1971
ZN	Zinc in the atmosphere (μg/m^3), 1971
MG	Manganese in the atmosphere (μg/m^3), 1971
PB	Lead in the atmosphere (μg/m^3), 1971
CD	Cadmium in the atmosphere (nanograms/m^3), 1968
CU	Copper in the atmosphere (μg/m^3), 1971
FE	Iron in the atmosphere (μg/m^3), 1971
CHEMINDX	Chemical index (sum of chemicals in the atmosphere
PESTPOIS	Pesticide poisoning risk, 1970 (risk per 1,000 children under fifteen years)
RATRISK	Rat bite risk, 1970 (risk per 1,000 children under fifteen years)

TABLE 25--Continued

Computer Code	Meaning
DISLAKE	Distance to Lake Michigan
DISH2O	Average distance to nearest body of water
OXYGEN[a]	Oxygen measuring parameters (biological oxygen demand--BOD)
MICRO[a]	Microbiological measuring parameter (fecal coliform)
ACID[a]	Acidity alkalinity measuring parameter (pH)
SOLIDS[a]	Total dissolved solids
CYANIDE	Total amount of cyanide (meet or exceed federal standards)
MBAS[a]	Methelyene blue active substances
OIL	Oil and grease (meet or exceed federal standards
PHENOLS	Phenols (meet or exceed federal standards)
CL	Chloride (meet or exceed federal standards)
FL	Fluoride (meet or exceed federal standards)
AMMONIA	Ammonia (meet or exceed federal standards)
NITRATE	Nitrates and nitrites (meet or exceed federal standards)
PHOS	Phosphorus (meet or exceed federal standards)
SULFATE	Sulfates (meet or exceed federal standards)
TRMETALS	Trace metals (number of violations/ten possible metals)
H2OINDX	Water index (number of violations/twenty-four parameters)
CUMINDX	Cumulative water index adjusted for distance to water (water index × distance to water)
DAYNOISE	Daytime background noise (dBA)
BACKNOIS	Background noise exceeding standards (dBA) in excess of the city standards)
INTER	Intermittent noise (dBA)
COMPLAIN	Noise complaints (actual number)
AIRPORT	Airport exposure (noise exposure forecast--NEF)
RES	Residential solid waste generation, 1972 (tons)

TABLE 25--Continued

Computer Code	Meaning
RATIO1	Residential generation/building, 1972 (tons)
RATIO2	Generation/residential dwelling unit, 1972 (tons)
RATIO3	Per capita residential solid waste generation, 1972 (tons)
COMM	Commercial solid waste generation, 1972 (tons)
RATIO4	Commercial generation/building, 1972
COMB	Combination solid waste generation, 1972 (tons)
IND	Industrial solid waste generation, 1972 (tons)
RATIO5	Industrial generation/building, 1972
BULK	Bulk trash, 1973 (tons)
CARS	Abandoned cars, 1973 (tons)
DIRT	Street dirt collected by mechanical sweeper, 1973 (tons)
TREES	Tree and stump removal, 1973 (tons of wood materials)
TEARDOWN	Residential domolition refuse, 1973 (tons)
SUBSID	Total subsidiary solid waste, 1973 (tons)
SW	Total solid waste generation (tons)
RATIO6	Total generation/building
TOTSW	Grand total solid waste generation (total solid waste generation plus subsidiary sources)
RATIO7	Per capita generation (tons/person/year)

[a]Eliminated from the analysis because they met the federal standards.

significant), but there was random ordering among the pairs so that there is no apparent trend (gamma=.17).

Three water pollution variables became important in the analysis: amounts of cyanide, fluoride, and phenols in the water. All three variables showed that where there were violations of this standard, the residents expressed high concern about pollution (gamma=-.67, p=.04). Where there were no violations, the attitudes were found to be negative and neutral (little or no concern).

In the case of noise pollution, only one variable was significant (p=.033)--exposure to airport noise. It was found that in areas with no exposure, attitudes were primarily positive or high concern (68%), followed by neutral (18%). However, in the exposure case, attitudes were found to be negative.

Only one solid waste pollution variable--generation of solid waste--showed a relationship to attitude about pollution. More than 40 percent of the areas with very high levels of total solid waste (without the subsidiary added in) also had positive attitudes about pollution (gamma=.75, p=.02). Low levels of concern are found exclusively in those areas with low levels of solid waste generation ($\leq 7,500$ tons/year). With medium levels of concern we also find low levels of waste generation. Medium to high levels of waste generation ($\geq 10,000$ tons/year) are associated with high levels of concern regarding pollution.

A closer examination of these relationships shows (table 26) that attitudes are negatively correlated with rat risk, distance from Lake Michigan, and airport noise.

The fact that decreasing distance from Lake Michigan was associated with high concern about pollution clearly indicates, as might be expected, that the closer one is to a body of water, the more likely he is to form an attitude about its being polluted. We found the highest concern levels in communities

TABLE 26

SPEARMAN RANK-ORDER CORRELATION MATRIX: ATTITUDE
TOWARD POLLUTION AND ENVIRONMENTAL QUALITY

Environmental Variable	Correlation Coefficient (r_s)
Ozone	.37*
Air index, 1975	.39*
Zinc	.46*
Cadmium	.49**
Pesticide poisoning	.40*
Rat risk	-.46*
Distance from Lake Michigan	-.58**
Ammonia	.39*
Airport noise	-.37*
Residential solid waste	.60**
Residential waste building	.54**
Residential waste dwelling unit	.35*
Commercial waste	.44*
Combination waste	.40*
Bulk waste	.43*
Abandoned cars	.41*
Street dirt	.52**
Residences demolished	.51**
Total subsidiary solid waste	.60**
Total solid waste generation	.68**
Total generation building	.47*
Grand total solid waste generation	.63**
Per capita generation	.53*

*$p < .05$.

**$p < .01$.

less than one mile from the lake; residents with the most negative attitudes lived more than five miles from the lake.

Table 26 also shows that high levels of solid waste, including the various sources (e.g., residential, commercial, industrial), are all positively associated with high concern about pollution. We find with some degree of significance ($p<.01$) that in areas where there are high levels of solid waste (all sources), there are also high concerns about pollution ($r_s=.68$) in that residents consider the pollution problem serious and are bothered by it.

Thus, some general statements may be made concerning attitudes toward pollution and environmental quality. Those areas of the city that rate very high on the air index (e.g., high pollution) have a high concern about pollution ($r_s=.68$) in that residents consider the pollution problem serious and are bothered by it.

Thus, some general statements may be made concerning attitudes toward pollution and environmental quality. Those areas of the city that rate very high on the air index (e.g., high pollution) have a high concern about pollution ($r_s=.39$, $p=.05$), although this association is not as strong as one might expect. Two trace metals found in the air--zinc and cadmium--are also positively associated with attitudes. Hence, areas of the city with high levels of zinc and cadmium in the atmosphere also have high concern scores. As has been mentioned, the only water pollution variable that influences attitudes is distance from Lake Michigan. In the noise category, the lower the exposure to airport noise, the higher the concern. Finally, we found that high solid waste levels indicated high concern about pollution. Areas with very high solid waste levels (regardless of the source) also had residents who considered pollution a serious problem, worried about it frequently, and were bothered by it.

Again, we must consider whether these same environmental variables show any association with attitudes toward pollution abatement or whether a different set of factors determines attitude toward pollution abatement.

Attitudes toward Pollution Abatement and Environmental Quality

In the cross-tabulation analysis, only one variable was found to be significantly ($p<.05$) associated with attitudes toward pollution abatement (AB). As with attitudes toward pollution, distance to Lake Michigan was again significant ($p=.04$). The gamma statistic ($-.34$) revealed that there was a mild negative trending between pairs, but no overall generalizations could be made. The largest percentage in the cell was 23 percent, for communities less than one mile from the lake that had negative attitudes and communities far from the lake that also had negative attitudes (18%). Thus, distance from Lake Michigan is not a factor in whether one is willing to do something about pollution and how effective one thinks that action will be.

Only eight variables were found to be factors in attitudes toward pollution abatement, using Spearman's rank-order correlation coefficient. All eight variables (table 27) were found to have a positive association with attitudes, although none of these were very strong ($r_s \leq .50$). High levels of ozone, trace metals in the atmosphere (particularly lead and iron), and a high air index were found in communities with positive attitudes toward abating pollution. Chloride in the water was also found to be positively associated with attitudes. Finally, two solid waste factors, bulk trash and street dirt, assumed significance. In areas where these factors were high, positive pollution abatement attitudes dominated. Again, we should note that none of these relationships are strong.

How well do these generalizations hold with regard to

TABLE 27

SPEARMAN RANK-ORDER CORRELATION: ATTITUDES TOWARD
POLLUTION ABATEMENT AND ENVIRONMENTAL QUALITY

Variable	Correlation Coefficient (r_s)
Ozone	.50**
Air index	.36*
Manganese in atmosphere	.36*
Lead in atmoshpere	.37*
Iron in atmosphere	.41*
Chloride	.42*
Bulk trash	.45*
Street dirt	.43*

*$p < .05$.

**$p < .01$.

attitudes about air pollution and air pollution variables, water pollution and water pollution variables, noise pollution and noise pollution variables, and solid waste pollution and solid waste pollution variables? Do the same generalizations made in attitudes toward general pollution abatement hold for the specific types of pollution? These questions are addressed in the following section.

Attitudes toward Specific Pollutants and Environmental Quality

Attitudes toward Air Pollution and Air Quality

Only one air parameter was found by cross-tabulation to be associated with attitudes toward air pollution--the amount of manganese in the air. Medium-low (0.06-1.0 µg/m^3) levels were associated with positive attitudes toward air pollution (gamma= .19, p=.048), although there was a more random ordering of the pairs, as indicated by the low gamma statistic.

On the other hand, Spearman's rank correlation test showed four variables that influenced attitudes toward air pollution. Carbon monoxide levels for both 1974 and 1975 were found to be negatively associated with attitudes toward air pollution ($r_s=-.47$, p=.014, and $r_s=-.30$, p=.09), although carbon monoxide for 1975 had less significance. Two trace metals in the atmosphere, zinc and cadmium, were found to be positively associated with attitudes toward air pollution ($r_s=.32$ and $r_s=.31$), but the significance of these correlations is not as high as one might wish--.071 for zinc and .085 for cadmium.

Attitudes toward Water Pollution and Water Quality

In the case of attitudes toward water pollution, only one variable assumed any significance in the contingency table analysis--the amount of ammonia in the water (whether or not federal standards were violated). It was found that 40.9 percent of the communities where there were no violations of the standards had negative attitudes about water pollution (gamma=.45, p=.062), but on closer examination of this association, ammonia was not found to be a significant factor. In the rank correlation matrix, the important variables were distance from Lake Michigan and amount of cyanide, phenols, and fluoride in the water. Distance from Lake Michigan was negatively associated with attitudes toward water pollution ($r_s=-.48$, p=.011). The remaining three water pollution parameters were also negatively associated with attitudes toward water pollution, although the associations were not as strong ($r_s=-.32$) or as significant (p=.071).

Attitudes toward Noise Pollution and Noise Levels

In the cross-tabulation analysis, no noise variables were significant. Airport noise was negatively correlated with attitudes toward noise ($r_s=-.30$, p=.05) in the Spearman rank correlation test. Background noise levels and intermittent noise levels were also negatively correlated ($r_s=-.24$ and $r_s=-.25$) with attitudes toward noise although with a lower significance level (p=.10).

Attitude toward Solid Waste Pollution and Solid Waste Levels

Five variables were found to be associated with attitudes toward solid waste pollution in the contingency table analysis. Three of these were significant at the .10 level (residential solid waste, combination solid waste, residential demolitions), and the remaining two (tree waste and total generation per building) were significant at the .05 level. In only one instance (the tonnage of residences demolished) was there more than a random ordering of the pairs. Low levels of residential demolition were associated (gamma=.52) with negative attitudes toward solid waste pollution. However, an examination of the Spearman correlation matrix shows twelve solid waste variables that have significant relations to attitudes toward solid waste pollution. As in the cross-tabulation analysis, the lower the tonnage of trees removed, the higher the concern about solid waste. Spearman's rho in this instance was -.32, with a significance level of .027. All the remaining eleven variables (table 28) had positive correlations. We find that the highest association is between attitudes toward solid waste pollution and the amount of subsidiary solid waste ($r_s=.50$). From this cursory look we can conclude that positive attitudes about solid waste levels are very high. The opposite of this is certainly also true.

Thus, we have provided the following findings on how attitudes toward specific pollutants relate to specific levels of pollution:

1. Attitudes toward air pollution are related to air

TABLE 28

SPEARMAN RANK-ORDER CORRELATION: ATTITUDES TOWARD
SOLID WASTE POLLUTION AND LEVELS OF
SOLID WASTE POLLUTION

Variable	Correlation Coefficient (r_s)
Residential generation	.33*
Commercial generation	.34*
Combination generation	.30*
Industrial generation building	.32*
Abandoned cars	.34*
Street dirt	.36*
Tree and stump removal	-.32*
Residential demolition refuse	.43**
Subsidiary solid waste	.49**
Total solid waste generation	.48**
All sources combined for grand total solid waste generation	.40**
Per capita generation	.38**

*$p < .05$.

**$p < .01$.

quality--positive attitudes (high concern) are found in areas with high levels of air pollution.

2. Attitudes toward water pollution are related to levels of water pollution (low levels are associated with negative attitudes), but they are also a function of the distance from the water.

3. High noise levels produce negative attitudes (people do not consider it a serious problem, nor are they bothered by it.

4. Areas with high levels of solid waste also have positive attitudes (high concern) about solid waste pollution.

It can be stated, then, that environmental quality does play some role in a community's attitude toward pollution--in particular toward air, water, and solid waste pollution. The higher the ambient levels of these parameters, the more likely the community is to consider the problem serious and to be bothered by it.

Do attitudes toward doing something about pollution show the same trends? For general pollution we found that the factors influencing attitudes toward the particular object were not necessarily the same factors that influenced attitudes toward doing something about pollution.

Attitudes toward Specific Pollution Abatement Strategies and Environmental Quality

Attitudes toward Air Pollution Abatement and Air Quality

Only one air variable was related to attitude toward air pollution abatement (ABA). Low and medium values on the air index were found to be associated with neutral ABA attitudes (gamma=.75, p=.069), but this was not the case in the correlation test. Here attitudes and the air index were positively related (r_s=.19) but not significant (p=.20). Utilizing the rank-order correlation procedure, four variables were found to have some association with attitudes toward air pollution. Ozone and trace metals in the air (manganese, lead, and cadmium) were positively associated with ABA, although these associations were generally on the weak side ($r_s \leq .39$) and significant only at the .05 level. Thus in areas with high levels of these pollutants in the air we find a prevailing positive attitude toward doing something about air pollution.

Attitudes toward Water Pollution Abatement and Water Quality

Six parameters were found to have significant associations

with attitudes toward water pollution abatement (ABW). Cyanide levels below the federal standards were found to have an association with negative and neutral ABW attitudes (gamma=.38, p=.004). This same pattern was found for oil (gamma=1.00, p=.001), phenols (gamma=.38, p=.004), chloride (gamma=1.00, p=.004) and fluoride (gamma=.38, p-.004). The cumulative water index was also significant, although the association of the pairs was much more random (gamma=.08, p=.001). Good water quality was associated first with negative ABW attitudes (27.3%), then with neutral ABW attitudes (22.7%). Fair water quality was associated with negative ABW attitudes (18.2%).

Oil and chloride appear significant on the rank-order correlation test. Both are positively associated with attitudes but have rather low correlation coefficients ($r_s=.36$ and $r_s=.40$, respectively).

The only difference in the two procedures was that distance to Lake Michigan again was important. It was negatively correlated with ABW, although this correlation was barely significant ($r_s=-.36$, p=.050).

One can conclude that in areas closest to the lake, positive attitudes prevail toward doing something about water pollution. Also, some measures of water quality (oil/grease and chloride) that exceed water quality standards are associated only mildly with attitudes toward water pollution abatement.

Attitudes toward Noise Pollution Abatement and Noise Levels

No significant variables relating these two factors were evident in the contingency table analysis or in the rank-order correlation analysis. Thus we conclude that noise levels have no bearing on community willingness to do something about noise pollution.

Attitudes toward Solid Waste Pollution Abatement and Solid Waste Levels

Only one variable was found to be associated with attitudes toward solid waste pollution abatement (ABSW)--industrial solid waste. Low levels of industrial solid waste were found to be associated with positive ABSW attitudes (gamma=-.74, p=.049); 68 percent of the communities fell into this cell in the contingency table. This also indicates that in those areas where there are high levels of industrial solid waste, the community is less willing (negative ABSW attitude) to do something about abating solid waste pollution.

This relationship does not appear significant in the rank-order correlation test (r_s=-.10, p=.293). The only factor that appears to be ranked with ABSW attitudes is the amount of bulk waste. Attitudes and bulk waste are mildly correlated (r_s=.36, p=.016), and high levels tend to be associated with positive attitudes toward doing something about solid waste pollution.

It appears from the foregoing analysis that, as an overall generalization, attitudes toward doing something about pollution are only mildly related to the actual pollution level. In the case of air pollution, high levels of ozone and some trace metals were found to be factors in determining attitudes, although they were not as strong as one would like. Also, only two water quality parameters were found to be correlated with attitudes toward abating water pollution. The strongest factor was distance from Lake Michigan, as it was with attitude toward water pollution.

Attitudes toward abating noise pollution were not influenced by noise levels, nor was there any correlation between noise levels and willingness to do something about the noise.

Finally, there was no real relationship between solid

waste levels and attitudes toward abatement. High levels of industrial waste were associated with negative attitudes, but there was no significant correlation. It appears that those areas with the lowest levels of industrial solid waste coincide with those that have the most positive attitudes about doing something about solid waste pollution.

Given these generalizations, is it possible to predict what a community might or might not do about pollution abatement? This is the focus of the next chapter.

CHAPTER VII

PREDICTION OF BEHAVIORAL INTENT AND BEHAVIOR

Prediction of human behavior has interested psychologists--in particular social psychologists--for many years, and geographers' interest in predicting behavior should be fairly obvious; for example, they need to know about acceptance of planning efforts by the public. Here we are interested in predicting behavior regarding pollution abatement. What kinds of actions are people willing to take to abate or alleviate pollution, and are these actions consistent with environmental planning efforts by local, state, or federal agencies? If an environmental planner knows what the residents of an area are willing to do and accept in terms of pollution abatement, it becomes easier to carry out such programs and ensure that the residents will accept them.

The assumption that has plagued social psychologists is that if a person likes some object, she should hold favorable beliefs about it and should intend to perform and thus actually perform favorable behaviors with respect to it. According to Fishbein and Ajzen (1975, p. 332), "the assumption of a close link between attitudes, beliefs, and intentions is justified only at a very global level." Empirical work on this topic has shown the inconsistency between a person's attitude toward an object and his specific intentions with respect to it. Thus it is apparent that one's attitude toward pollution, for example, would not necessarily be related to one's intent to do something about it.

Given this dichotomy, Fishbein and Ajzen have developed their theory of behavioral intention. This theory (see chapter

3) specifies the immediate determinants of an individual's intention to engage in a given behavior and thus provides the means for predicting such intentions (p. 332). A person's intention to perform a given behavior is a function of two types of determinants: attitudinal (attitude toward performing the behavior in question); and normative (belief that the relevant referents--in this case society--think she should or should not perform the behavior in question, and motivation to comply with the referents).

We are interested in two intentions--the intention to abate pollution in general and the intention to abate solid waste pollution. We will first establish the methodology with the general case, then see whether the specific case conforms to the same model.

Prediction of Behavioral Intent to Abate Pollution

Utilizing Fishbein's model, we find that behavioral intent is a function of attitudes toward specific pollution abatement measures and subjective norms (eq. 7.1).

$$B \approx BI = AB_{w_1} + SN_{w_2}, \qquad (7.1)$$

where B = behavior   AB = attitude toward the behavior
       BI = behavioral intent SN = subjective norms
                               $w_1 w_2$ = empirical weights.

The weights in this case indicate the relative importance of specific factors (or independent variables) in the prediction of intent. The standard least-squares regression model was used, and the regression coefficients were used as the weights. Beta coefficients were utilized in discovering which of the two factors was more important in predicting intent (i.e., how much each factor influenced prediction).

Before we proceed with the first regression equation,

let us look at the simple correlations among the attitude variables (table 29). Here we find high correlations between intent and attitudes toward pollution abatement, between intent and subjective norms, and between attitude toward pollution abatement and subjective norms. Thus there is an apparent problem of collinearity between our two supposedly independent factors in the prediction of intent. However, as Rao and Miller (1971, p. 48) state, "One should realize that simple correlations are only elements of the entire correlation matrix and hence, may or may not contribute to problems of multicollinearity. <u>One should not, a priori, rule out estimation of any regression equation because of high simple correlations between any two independent variables.</u>" We also found that there were relatively low correlations between intent and overt behavior. One aspect to examine is the correlations between attitudes toward pollution and the other behavioral components. We find that attitudes toward pollution (AO) and attitudes toward pollution abatement (AB) are slightly correlated for the city as a whole. But in three communities we find higher correlations. In North Lawndale and Roseland we find that AO and AB are positively correlated (r=.66 and r=.16, respectively), whereas in Woodlawn we find AO and AB negatively correlated (r=-.74). Attitudes toward pollution and intent to abate pollution are also found to be only slightly correlated. Attitudes toward pollution and overt behavior are not correlated. Whether these generalizations hold for each of the eight types and then for each of the communities studied will be discussed in the following section when we look at differences between area types and between communities.

What are the results of the first regression equation? We found that both attitudes toward pollution abatement and subjective norms were significant in predicting intent (table 30),

TABLE 29

PEARSON PRODUCT-MOMENT CORRELATION COEFFICIENTS

Area	AO-AB	AO-SN	SN-I	AO-B	AB-SN	AB-I	AB-B	SN-I	SN-B	I-B
Chicago	.36**	.36**	.39**	.24**	.73**	.95**	.26**	.74**	.30**	.30**
Type 1 HP, LSES, W										
Uptown	.36**	.40**	.40**	.36**	.67**	.95**	.30**	.68**	.43**	.37**
Type 2 HP, LSES, B										
West Garfield Park	.31	.32	.39*	.31	.83**	.94**	.31	.86**	.38*	.48**
East Garfield Park	.44*	.47*	.46*	.42*	.72**	.86**	.32	.85**	.49*	.42*
North Lawndale	.66**	.38	.64**	.08[a]	.62**	.96**	.07[a]	.67**	.04[a]	-.03[a]
Woodlawn	-.74**	-.64*	-.74**	---	.81**	.79**	---	.78**	---	---
Englewood	.45**	.48**	.42**	-.08	.75**	.91**	.22	.79**	.24	.16
Type 3 LP, LSES, W										
Burnside	.47**	.44**	.54**	.33*	.78**	.97**	.47**	.76**	.46**	.47**
Pullman	.41**	.38**	.49**	.30**	.68**	.95**	.43**	.66**	.36*	.47**
Type 4 LP, LSES, B										
Avalon Park	.29**	.28**	.35**	.29**	.71**	.91**	.28**	.74**	.33**	.27**
Type 5 HP, HSES, W										
Lincoln Park	.51**	.74**	.56**	.15	.68**	.95**	.36**	.76**	.36*	.27
Near North	.50**	.55**	.49**	.36**	.78**	.95**	.28**	.81**	.30*	.38**
Hyde Park	.43**	.37**	.44**	.39**	.66**	.79**	.41**	.69**	.32*	.44**

TABLE 29--Continued

Area	AO-AB	AO-SN	SN-I	AO-B	AB-SN	AB-I	AB-B	SN-I	SNOB	I-B
Type 6 HP, HSES, B										
Kenwood	.41**	.24*	.42**	.30**	.63**	.93**	.22*	.71**	.16	.26*
South Shore	.28*	.20	.30*	.26*	.66*	.94**	.22*	.70**	.35	.22
Type 7 LP, HSES, W										
Clearing	.55**	.48*	.49*	.28	.86**	.98**	.54**	.86**	.33	.48*
Garfield Ridge	.11	.01	.08	-.23	.89**	.94**	.12	.92**	.19	.23
Dunning	.09	.20	.08	.16	.91**	.87**	.21	.78**	.32	.19
Montclare	.56**	.56**	.55**	.38*	.84**	.94**	.51**	.83**	.46**	.54**
Forest Glen	-.19	-.23	-.17	.04	.78**	.94**	.36	.79**	.32	.22
Beverly	.49**	.57**	.44**	.41*	.85**	.97**	.62**	.83**	.64**	.63**
Type 8 LP, HSES, B										
Roseland	.61**	.50**	.62**	.51**	.78**	.79**	.48**	.79**	.47**	.52**
Calumet Hieghts	.36**	.31**	.38**	.05	.73**	.92**	.18**	.76 *	.22**	.17

*p < .01.

**p < .05.

[a]Coefficient cannot be computed (N=10).

TABLE 30

LEAST-SQUARES REGRESSION OF INTENT TO ABATE POLLUTION
ON ATTITUDES TOWARD ABATEMENT AND SUBJECTIVE NORMS

Area	Constant	Multiple Regression Coefficients		Beta Coefficients		R^2
		AB	SN	AB	SN	
Chicago	5.70	.086** (3221.47)[a]	.058** (64.56)	.857	.121	.90
Type 1 (HP, LSES, W)	5.58	.098** (625.78)	.037* (3.89)	.905	.071	.91
Type 2 (HP, LSES, B)	5.92	.074** (268.33)	.082** (17.42)	.788	.201	.90
Type 3 (LP, LSES, W)	5.55	.010** (426.20)	.009 (0.21)	.943	.021	.92
Type 4 (LP, LSES, B)	5.60	.079** (218.41)	.096** (13.65)	.773	.193	.84
Type 5 (HP, HSES, W)	5.83	.088** (629.24)	.061** (15.84)	.862	.137	.93
Type 6 (HP, HSES, B)	5.74	.080** (361.46)	.086** (16.72)	.819	.176	.89
Type 7 (LP, HSES, W)	5.49	.087** (222.46)	.061* (4.19)	.840	.115	.88
Type 8 (LP, HSES, B)	6.00	.081** (474.68)	.062** (9.83)	.859	.124	.92

*$p < .01$.

**$p < .05$.

[a]Numbers in parentheses are the F-statistic.

with an adjusted $r^2=.897$. Attitudes were more important for predicting intent for Chicago residents as a whole ($\beta=.86$) than were subjective norms ($\beta=.12$). Thus we find for the city as a whole this general behavioral-intention model:

$$BI = 5.70 + (.086AB) + (.058SN) \qquad (7.2)$$

When the area types are individually examined, we find only one instance (type 3) where one independent variable was not significant. For type 3 (low pollution, low status, white) the overriding factor in predicting intent to abate pollution was attitude toward actions of abatement. We also found that in types 2, 4, 5, and 6 subjective norms played more of a role than for the city as a whole, but that attitudes were still the most important factor.

At the community level, the following results are evident (table 31). For half the communities, attitudes were the only significant factor in predicting behavioral intent. And in the low pollution, low status, black type (Avalon Park) and the low status, low pollution, white type (Burnside and Pullman), attitudes toward pollution abatement were the only factors significant in predicting intent. The other communities were dispersed throughout the remaining types, and there did not appear to be any particular trend. But we did find that in two area types (high pollution, high status, white, and high pollution, high status, black), encompassing Lincoln Park, Near North, Hyde Park, Kenwood, and South Shore, both attitudes and subjective norms were important.

In Garfield Ridge, subjective norms were more important in predicting intent than they were in the city as a whole, although not as important as attitudes ($\beta=.38$ and $\beta=.60$, respectively). We also found that the subjective norms played a major role in intentions in West Garfield Park ($\beta=.24$) and

TABLE 31

LEAST-SQUARES REGRESSION OF INTENT TO ABATE POLLUTION ON ATTITUDES TOWARD POLLUTION ABATEMENT AND SUBJECTIVE NORMS

Area	Constant	Multiple Regression Coefficients		Beta Coefficients		R^2
		AB	SN	AB	SN	
Uptown	5.58	.098**	.037**	.905	.071	.91
West Garfield Park	5.65	.069**	.080*	.747	.236	.90
East Garfield Park	6.26	.059**	.150**	.530	.465	.83
North Lawndale	5.68	.086**	.057	.896	.109	.93
Woodlawn	5.78	.078**	-.017	.993	-.030	.92
Englewood	6.32	.064**	.092*	.729	.245	.85
Burnside	5.62	.102**	-.000	.970	.001	.94
Pullman	5.52	.098**	.010	.932	.021	.89
Avalon Park	5.60	.079**	.096**	.773	.193	.84
Lincoln Park	5.80	.082**	.072**	.810	.207	.92
Near North	6.09	.085**	.080*	.823	.166	.91
Hyde Park	5.67	.094**	.046*	.913	.087	.94
Kenwood	5.54	.079**	.105**	.807	.196	.88
South Shore	6.05	.080**	.063*	.843	.145	.89
Clearing	5.58	.081**	.042	.904	.086	.96
Garfield Ridge	6.03	.067**	.173*	.596	.384	.90
Dunning	5.51	.094**	-.022	.902	-.037	.73
Montclare	5.28	.090**	.071	.834	.129	.88
FOrest Glen	5.18	.091**	.083	.839	.131	.88
Beverly	5.45	.092**	.022	.929	.046	.93
Roseland	5.74	.086**	.044	.904	.082	.94
Calumet Heights	6.61	.069**	.080**	.774	.174	.85

*p < .05.

**p < .01.

Englewood ($\beta=-.24$). The most exciting finding was for East Garfield Park, where subject norms and attitudes were of approximately equal importance in predicting intent ($\beta=.46$ and $\beta=.53$, respectively). Finally, attitudes toward pollution abatement accounted for 99 percent of Woodlawn residents' intention to abate pollution.

In some areas there are very strong feelings of peer pressure, which might take a variety of forms including pride in the community, conformity, and wondering what neighbors would think. We suspect that this strong feeling would be found in highly ethnic areas owing to their strong cultural bias. Instead, we find high peer-pressure scores (subjective norms) in the low status black communities.

Prediction of Behavioral Intent to Abate Solid Waste Pollution

In examining the simple correlations among the components of attitude toward solid waste (table 32), we find much less of a collinearity problem. We do, however, find very high correlations between attitudes about pollution abatement actions and intentions relating to pollution abatement. In comparing the two attitudes (AO and AB), we find that they are strongly correlated in only two communities--North Lawndale, where they are positively correlated ($r=.77$, $p=.01$), and Woodlawn, where they are negatively correlated ($r=-.62$, $p=.05$). Only in Montclare are there high correlations between intent to perform a behavior and actual performance of that behavior. As with general pollution, there is no apparent correlation between what a community intends to do about pollution and what residents actually do.

In comparing general pollution with solid waste pollution, we found that both attitudes and societal norms contributed to predicting intent (table 33). As before, attitudes played the

TABLE 32

PEARSON PRODUCT-MOMENT CORRELATION COEFFICIENTS: SOLID WASTE

Area	AO-AB	AO-SN	AO-I	AO-B	AB-SN	AB-I	AB-B	SN-I	SN-B	I-B
Chicago	.21**	.30**	.26**	.13**	.58**	.92**	.29**	.60**	.26**	.31**
Uptown	.35**	.35**	.38**	.26**	.56**	.93**	.39**	.57**	.40**	.43**
West Garfield Park	.34*	.54**	.36*	.11	.58**	.98**	.48**	.56**	.25	.52**
East Garfield Park	.02	.35	.16	.12	.49*	.89**	.08	.60**	.15	.26
North Lawndale	.77**	.41*	.75**	.20	.63**	.96**	.08	.57**	.26	.03
Woodlawn	-.62*	-.16	-.50	-.69*	-.05	.97**	.53	-.01	.24	.48
Englewood	.40*	.63**	.39*	-.12	.62**	.86**	.13	.64**	.14	.18
Burnside	.31*	.46**	.37*	.44**	.62**	.93**	.34*	.60*	.50**	.45**
Pullman	.28*	.20	.35**	.22	.57**	.92**	.46**	.60**	.26**	.46**
Avalon Park	.20*	.29**	.30**	.10	.50**	.92**	.25**	.56**	.33**	.29**
Lincoln Park	.50**	.68**	.46**	.07	.60**	.93**	.21	.58**	.20	.23
Near North	.08	.23	.17	.15	.48**	.87**	.11	.52**	-.00	.14
Hyde Park	.40**	.31*	.35**	.19	.61**	.95**	.47**	.59**	.40**	.49**
Kenwood	-.01	-.07	-.03	.18	.67**	.93**	.32**	.70**	.06	.29*
South Shore	-.03	.11	.02	.04	.44**	.88**	.39**	.47**	.24	.40**

TABLE 32--Continued

Area	AO-AB	AO-SN	AO-I	AO-B	AB-SN	AB-I	AB-B	SN-I	SN-B	I-B
Clearing	.46*	.52**	.42*	.29	.84**	.97**	.44*	.82**	.29	.47*
Garfield Ridge	.39*	.21	.38*	.12	.77**	.93**	.29	.81**	.44*	.32
Dunning	.03	-.04	-.12	-.13	.63**	.84**	-.00	.74**	.16	.13
Montclare	.45*	.56**	.43*	.57**	.49**	.93**	.58**	.58**	.37*	.59**
Forest Glen	.07	.14	-.05	.28	.63**	.96**	.35	.62**	.45*	.33
Beverly	-.07	.18	.02	.40	.72**	.93**	.40*	.83**	.36*	.43*
Roseland	.31**	.35**	.35**	.31**	.80**	.92**	.28**	.82**	.30**	.24*
Calumet Heights	.08	.30**	.10	.00	.55**	.82**	.23*	.60**	.10	.08

*p < .05.

**p < .01.

TABLE 33

LEAST-SQUARES REGRESSION OF INTENT TO ABATE SOLID WASTE POLLUTION ON ATTITUDES TOWARD SOLID WASTE ABATEMENT AND SUBJECTIVE NORMS

Area	Constant	Multiple Regression Coefficients		Beta Coefficients		R^2
		ABSW	SNSW	ABSW	SNSW	
Chicago	4.30	.076** (2944.70)[a]	.044** (51.82)	.851	.113	.85
Type 1 (HP, LSES, W)	4.20	.085** (486.88)	.024* (3.09)	.888	.071	.86
Type 2 (HP, LSES, B)	4.65	.072** (389.80)	.031* (4.54)	.874	.094	.85
Type 3 (LP, LSES, W)	4.21	.079** (290.19)	.031 (2.35)	.874	.079	.85
Type 4 (LP, LSES, B)	4.04	.078** (408.97)	.057** (10.67)	.847	.137	.85
Type 5 (HP, HSES, W)	4.15	.081** (456.42)	.030* (3.39)	.881	.076	.85
Type 6 (HP, HSES, B)	4.34	.072** (298.64)	.048** (6.49)	.837	.123	.83
Type 7 (LP, HSES, W)	4.16	.071** (400.49)	.081** (18.86)	.812	.176	.88
Type 8 (LP, HSES, B)	4.70	.064** (221.25)	.085** (15.27)	.756	.199	.82

*p < .05.

**p < .01.

[a] Numbers in parenthesis are the F-statistic.

major role, accounting for 85 percent of the variance, while subjective norms played a much more minor role ($\beta=.113$). This is very similar to the general case, as illustrated below (see eq. 7.2 for the general case):

$$BISW = 4.30 + (.076ABSW) + (.044SNSW) \qquad (7.3)$$

When we look at the individual types, we again find one instance where neither independent variable is a factor. This is type 3, where it appears that attitudes toward pollution abatement are the overriding influence in predicting intent. We also found that for four of the types (1, 2, 3, 5) social norms were much less of an influence than for the city as a whole, and that in the remaining communities social norms were more influential than for the city as a whole.

When the individual communities are examined, only eight show both factors as significant (table 34). Whether this is in fact due to collinearity problems between the independent variables or just the dominance of the attitude component is not clear at this time. In four of the communities where both factors are important in predicting intent, we find that the subjective norm component is substantially higher than for the city as a whole. For example, in Dunning and Beverly, both low polluted, high status, white areas, we find that subjective norms account for .356 and .328, respectively, of the influence in the prediction. We also find that in the low polluted, high status, black areas (Roseland and Calumet Heights), subjective norms acquire more importance than for the city as a whole ($\beta=.241$ and $\beta=.206$, respectively).

What can be said is that attitudes toward performing pollution abatement behavior are the overriding factor in predicting intent to perform that behavior. In some cases, subjective norms play a role, but only on a minor level. This

TABLE 34

LEAST-SQUARES REGRESSION OF INTENT TO ABATE SOLID WASTE POLLUTION
ON ATTITUDES TOWARD SOLID WASTE ABATEMENT AND SUBJECTIVE NORMS
(By Area)

Area	Constant	Multiple Regression Coefficients		Beta Coefficients		R^2
		ABSW	SNSW	ABSW	SNSW	
Uptown	4.30	.076**	.044**	.851	.113	.85
West Garfield Park	3.84	.094**	-.003	.984	-.008	.96
East Garfield Park	4.84	.780**	.060	.780	.219	.80
North Lawndale	4.79	.080**	-.020	.999	-.060	.92
Woodlawn	4.60	.071**	.020	.972	.043	.92
Englewood	5.44	.050**	.048	.763	.159	.74
Burnside	4.04	.082**	.020	.898	.049	.85
Pullman	4.30	.076**	.045*	.850	.115	.84
Avalon Park	4.04	.078**	.057**	.850	.137	.85
Lincoln Park	4.08	.086**	.012	.908	.036	.86
Near North	4.01	.082**	.047	.809	.138	.77
Hyde Park	4.28	.081**	.008	.942	.014	.90
Kenwood	4.03	.076**	.066**	.832	.142	.87
South Shore	4.80	.065**	.032	.829	.109	.77
Clearing	3.68	.091**	-.005	.981	-.013	.93
Garfield Ridge	4.65	.058**	.095	.755	.228	.88
Dunning	4.39	.058**	.163**	.622	.356	.77
Montclare	4.13	.072**	.073**	.842	.175	.87
Forest Glen	3.54	.092**	.011	.952	.018	.92
Beverly	4.45	.055**	.160**	.692	.328	.90
Roseland	4.52	.063**	.122**	.729	.241	.86
Calumet Heights	5.24	.056**	.066**	.707	.206	.69

*p < .01.
**p < .05.

appears as somewhat contrary to the theory as put forth by Fishbein and Ajzen, but there are many possible explanations for this. One explanation is that we were not dealing with a controlled laboratory situation where we could actually observe individuals. We relied on secondhand information--questionnaire responses. It is possible that people have a tendency to lie on these questionnaires to place themselves in a better light or simply to respond as they think the interviewer wants them to.

There may also have been problems with the questionnaire itself. Psychologists prefer a bipolar scale on which the respondent merely places a check where he feel his response lies. On the telephone one does not have this luxury, and the range of responses becomes more limited. Also, it is apparent that the wording of the intent question and the computation of the variable on attitude toward abating pollution might have been in error. That is, calculating two different measures from basically the same question may have in fact accounted for the high degree of collinearity between the AB and I measures.

The correlation between intent and behavior was generally not very high, and this leads to questioning the theory that implies that intent is approximately equivalent to behavior. It would be interesting to examine the relationship between these two factors in order to construct a model that ultimately predicted behavior. This is beyond the scope of the present research, but it is not beyond future investigation, especially the kind of relation that exists (e.g., whether linear or curvilinear) and whether an attitudinal and a normative component would be the only factors in the prediction.

CHAPTER VIII

CONCLUSIONS AND OPPORTUNITIES FOR FUTURE RESEARCH

Let us briefly review the questions that provided the main focus of this research.

1. Do social characteristics account for differences in attitudes toward pollution?

2. Do attitudes toward pollution vary by level of exposure to pollution?

3. Can differences in attitude toward pollution abatement be distinguished by social characteristics?

4. Are attitudes toward pollution abatement a function of the level of pollution for the area?

The following conclusions can be drawn from the preceding analysis.

1. Community attitudes can be differentiated by social characteristics. (a) Persons in white-collar occupations are less concerned and feel less bothered by pollution than blue-collar workers. (b) Blacks are likely to feel that pollution is more serious and bothersome than whites are. (c) Apartment-dwellers are more bothered by pollution and find it a more serious problem than homeowners do. (d) Persons in high-valued homes exhibit less concern about pollution than do those who live in homes that are low-valued.

2. Pollution abatement findings showed that the white-collar black population and the socially depressed black areas had the most positive attitude toward doing something about pollution. (a) Persons living in areas with high air pollution levels were more bothered than people living in areas with lower

air pollution levels. (b) Distance from Lake Michigan was the most important factor in determining attitudes toward water pollution. (c) High levels of solid waste were found in those areas that also had a very positive attitude toward pollution (i.e., they considered the problem very serious, were bothered by it, and often worried about it). (d) The level of noise pollution was not a factor in attitudes toward noise pollution.

In reference to the questions posed at the beginning of the chapter we find:

1. Differences in attitude can be accounted for by differences in social characteristics.

2. Communities with high levels of pollution tend to consider pollution a serious problem, to worry about it often, and to be bothered by it in some way.

3. Attitudes toward pollution abatement can also be distinguished on the basis of social characteristics, with blacks showing more positive attitudes about doing something to abate pollution than whites.

4. A community's willingness to do something about pollution is not a function of the pollution levels for that area.

As a secondary focus we wanted to determine whether we could predict a person's behavior or her intent to perform a given behavior assuming we knew her attitude and had some measure of social norms. Also, we had some interest in seeing which of the two factors was the most important or carried the most weight in predicting intent: attitude or social norms. It was found in an analysis that the overriding factor was attitude, and which in all cases overshadowed the social norms component. Whether this is a function of the questionnaire information or the calculation of the variables is open to question at this point and will need to be resolved by future investigation.

Many other questions that arose during the course of this investigation remain unresolved and provide avenues for future research.

1. Is intent really approximately equal to actual behavior? According to the theory this is so, but our results are not conclusive, since there are very low correlations between the intent measures and the behavior measures.

2. Are the generalizations concerning attitudes and social differentiation valid for another metropolitan area, or is Chicago unique? Since this particular methodology has never been employed in previous attitude studies, it appears that some aspect of regional differences might be an important determinant of attitude. As even the scant literature shows, there is still some disagreement on whether social differences are important.

3. Do abatement attitudes vary by regions and areas of the country? There has been no comprehensive study of abatement strategies that might show whether the conclusions we found for Chicago would be applicable to other areas.

4. Can behavior be predicted using a different version of the Fishbein model? Here the concern would be with the lack of prediction power in the equation as we used it for Chicago. It appears that some formulation utilizing intent as well as attitudes toward the particular object--in this case pollution--might prove a better predictor of actual behavior.

5. Do social differences in the respondents account for the lack of precision in the behavioral intent model in terms of predicting behavior? The main focus again would be on improving the capabilities of the model in simultaneously controlling for social differences among the respondents.

These are only a few of the many questions raised by

this research. We hope that answers to these specific questions will be forthcoming shortly, as well as opportunities to apply this type of analysis to policy formulation.

APPENDIXES

APPENDIX 1

LITERATURE REVIEW

Study	Town	Years	Sample Population	Sample Size	Type of Pollution	Questionnaire Type	Concepts Measured	S.E.S. and Perception	Exposure and Perception
Smith, Zeidberg, and Schuenemann 1964	Nashville	1958-59	Households	2,835	Air	Personal	Concern	*	*
Medalia 1964	Nashville	1961-62	Households	104	Air	Personal	Awareness, concern, action potential	*	--
Medalia and Finkner 1965	Clarkston, Wa.	1962	Households	100	Air	Personal	Attitudes, concern	*	--
Southern Illinois University 1965	Saint Louis	1963	Households	1,000	Air	Telephone	Awareness, knowledge, acceptance of control	--	--
de Groot et al 1966	Buffalo	1959, 1962	Population	800	Air	Telephone, personal	Seriousness of problem, concern, levels of awareness	*	*
Williams and Burnyard 1966	Saint Louis	--	Households, public officials	1,000	Air	--	Opinions, concern	--	*
Shusky 1966	Saint Louis	--	Household	1,002	Air	--	Attitudes, definition, concern	--	--

Continued

* = yes; -- = no.

APPENDIX 1--Continued

Study	Town	Years	Sample Population	Sample Size	Type of Pollution	Question-naire Type	Concepts Measured	S.E.S. and Perception	Exposure and Perception
Stalker and Robinson 1967	Birmingham, U.K.	--	Households	7,200	Air	--	Opinion	--	--
Johnson 1968	Northeast Illinois	1954-64	Complaints	190	Air	None	All aspects of complaints	*	--
Crowe 1968	Johnstown, Pa.	1965	Population	600	Air	Personal	Definition	*	--
Rankin 1969	Charleston, W.Va.	--	Population	1,400	Air	Personal	Attitudes	*	--
Creer, Gray, and Treshow 1970	Utah	1970	Residents	110	Air	Personal	Concern for abatement	*	--
Auliciems and Burton 1971	Toronto	1967, 1969	Residents	414	Air	Personal	Awareness	*	--
Swan 1972	Detroit	1969	High-school youths	173	Air	Personal, game simulation	Awareness, concern, knowledge	*	--
Johnson 1972	Pomona, Costa Mesa Claremont, Calif.	1970	Households	122	Air	Personal	Responsibility, interrelatedness with other problems	*	--
Kromm 1973	Yugoslavia	1970	Households	168	Air	Personal	Awareness, coping responses	*	--

154

Author	Location	Year	Sample	N	Topic	Scale	Variables	
Kromm, Probald, and Wall 1973	U.K., Budapest, and Yugoslavia	1971	Residents	400	Air	Personal	Awareness, coping strategies	*
Wall 1973a	South Yorkshire, U.K.	1972	Residents	120	Air	Personal	Definitions, control efforts, magnitude	--
Wall 1973b	Sheffield, U.K.	1972	Residents	120	Air	Personal	Concern, control efforts	*
Barker 1974	Toronto		Students		Air	Personal	Definition, concern, knowledge, social responsibility, alternatives to problem	*
Unwin and Holtby 1974	Manchester, U.K.	1970	Population	100	Air	Personal	Awareness of control, adaptation of control	--
Johnston and Hay 1974	Christchurch, New Zealand	1973	Population	150	Air	--	Awareness	--
Dasgupta 1967	Mississippi		Local landowners	349	Watershed development	Personal	Attitudes	*
Frederickson and Magnas 1968	Syracuse	1967	Population	1,036	Water	Personal	Priorities, attitudes	--

Continued

APPENDIX 1--Continued

Study	Town	Years	Sample Population	Sample Size	Type of Pollution	Question-aire Type	Concepts Measured	S.E.S. and Perception	Exposure and Perception
Dillehay, Bruvold, and Siegel 1969	California	--	Residents	145	Drinking water	Personal	Favorability	--	--
Ibsen and Ballweg 1969	Virginia	--	Population	453	Water resources	Telephone	Concern, action taken, responsibility	--	--
Mitchell 1971	Waterloo, Canada	--	Population, water managers	400	Water resources	Personal	Priorities	--	--
O'Riordan 1971	British Columbia	--	Population	266	Lake Pollution	Personal	Concern, awareness, actions	--	--
David 1971	Wisconsin	1969	Population	574	Water	Personal	Definition, indicators	*	*
Bruvold 1971	San Francisco, San Diego	1969	Population	50	Water	Personal	Attitudes	--	--
Dynes and Wenger 1971	--	--	Decision-makers	--	Water problems	Personal	Seriousness	--	--
Costantini and Hanf 1972	Lake Tahoe	1969-70	Decision-makers	303	Water	Personal	Concern	--	--

Ballweg 1972	Roanoke, Va.	1971	Households	307	Dual use of water resources	Personal	Attitudes	*	--
Sims and Baumann 1974	Kokomo, Ind. Lubbock, Tex. San Angelo, Tex. Colorado Springs Santee, Calif.	--	Population	2,000	Renovated waste water	Personal	Responses to use of renovated waste water	*	--
Watkins 1974	West Palm Beach and Homestead, Fla.	--	Population	313	Water use	Personal	Attitude	*	--
Ditton and Goodale 1974	Green Bay, Wisc.	1971	Households	2,174	Recreation and water quality	Personal	Usage, improvement measures, description of quality	--	*
Gore, Wilson, and Capener 1975	Fall Creek, N.Y.	1971	Households	623	Water problems	Personal	Concern--dimensions and sources, awareness, abatement measures	*	--
Committee 1963a	London	1961-62	Population	1,400	Noise	Personal	Annoyance, sources	--	--
Committee 1963b	Heathrow Airport, London	1961	Population	1,909	Aircraft noise	Personal	Concern, tolerability, adaptability, physiological and behavioral effects of noise	--	--

Continued

APPENDIX 1--Continued

Study	Town	Years	Sample Population	Sample Size	Type of Pollution	Question-naire Type	Concepts Measured	S.E.S. and Perception	Exposure and Perception
Robinson, Bowsher, and Copeland 1963	Farnborough air show, U.K.	1961	Volunteers	60	Aircraft and flight-generated noise	Experimental	Annoyance	*	--
Nixon and Borsky 1966	Saint Louis	1961-62	Households	1,000	Sonic booms	Personal, experimental	Annoyance	--	--
Bishop 1966	Los Angeles Int'l Airport	1965	Subjects	55	Aircraft	Experimental	Absolute judgments of noise level, relative judgments of noise level	*	*
Bolt, Beranek, and Newman 1967	Los Angeles, Boston, and New York	1966-67	Households	259	Noise	Personal	Concern, annoyance, source	*	--
Griffiths and Langdon 1968	London	1967	Population	1,200	Road traffic	Personal	Attitudes, opinions	--	--
Jonsson et al. 1969	Ferrara, Italy, and Stockholm	--	Population	366	Vehicular traffic noise	Personal	Annoyance	--	--

Continued

Bragdon 1970	Philadelphia	--	Population	500	Community noise	Personal	Response to noise concern	*	*
McKennell 1970	--	1961, 1965	Population	--	Aircraft noise	Comparison	Annoyance, complaints	*	*
Cameron, Robertson, and Zaks 1972	Detroit and Los Angeles	1968	Population	2,626	Noise and sound pollution	Telephone	Relation of exposure to illness, exposure	*	*
Koczkur et al. 1972	Toronto	1971	Population	7,614	Noise	Personal	Concern, annoyance	--	*
Klee 1971	Ten cities	1969	Households	1,609	Sanitary landfill	Personal	Awareness, lattitude of acceptance, rejection, and non-commitment	--	--
National Analysts 1973	Six cities	1971-72	Housewives	1,282	Solid waste	Personal	Knowledge, acceptability, attitudes, current practices	--	--
Sigler 1973	Illinois	1972	Population	3,597	Solid waste	Telephone	Awareness, knowledge of disposal services, attitudes	*	--

Continued

APPENDIX 1--Continued

Study	Town	Years	Sample Population	Sample Size	Type of Pollution	Questionnaire Type	Concepts Measured	S.E.S. and Perception	Exposure and Perception
Finnie 1973	Philadelphia	1969	Population	--	Litter	Observed, experimental	Rate of littering	*	*
Jonsson 1963	Sweden	1956-61	Residents	257	Air and noise	Personal	Annoyance, reactions to stimuli	*	*
Van Arsdol, Sabagh, and Alexander 1964	Los Angeles	1961-62	Residents	824	Smog, air traffic noise, brush fire, flood, earth slide	Personal	Seriousness	*	*
Lycan and Sewell 1968	Victoria, B.C.	1966	Population	120	Water and air	Personal	Attitudes, coping strategies	*	--
McMeiken and Rostron 1970	Courtenay, Victoria, B.C.	1968	High-school seniors	300	Air and water	Personal	Concern	--	--
Cooke and Saarinen 1971	Tucson	1970	Population	217	General	Personal	Seriousness, levels of satisfaction, sources, responsibility, solutions	*	--

Winham 1972	Hamilton, Ont.	1970-71	Population	685	General	Personal	Growth vs. pollution attitudes, concerns vs. operational attitudes	*	--
Jacoby 1972	Detroit	1969	Population	506	Air, water, noise	Personal	Attitudes, concern	*	*
Tognacci et al. 1972	Boulder, Colo.	--	Population	141	General	Personal	Concern, beliefs, willingness to take personal action, evaluation of anti-pollution movement	*	--
Rich 1974	Ebensburg, Pa.	1973	Population	128	General	Personal	Perception as related to political response	--	--

APPENDIX 2

QUESTIONNAIRE

COMMUNITY AREA NO. ___ ___ ___ ___ 1-4/

TELEPHONE NUMBER _ _ _ _ _   _ _ _ _ 5-11/

RESPONDENT'S SEX: Male........ 1   12/
                    Female...... 2

POLLUTION SURVEY

Hello, I'm _____ calling from Telesurveys of Illinois. We are conducting a survey for the University of Chicago's Geography Department to learn how people of Chicago feel about problems in our city. Your name was selected at random, and all of the information will be treated confidentially. The total interview will take less than 10 minutes.

As someone who lives in the city, you are faced with a number of concerns and a lot of problems.....

1. In the last month how often have you thought or worried about the following problems? How about crime. Would you say you thought or worried about crime very often, from time to time, or not much at all? REPEAT EACH CATEGORY AS NEEDED

	Very often	From time to time	Not much at all	NA/DK	
a. Crime...........................	1	2	3	4	13/
b. High taxes......................	1	2	3	4	14/
c. Decay of your neighborhood.......	1	2	3	4	15/
d. Drugs and narcotics..............	1	2	3	4	16/
e. Racial problems..................	1	2	3	4	17/
f. Pollution.......................	1	2	3	4	18/
g. Reliability of transportation....	1	2	3	4	19/
h. Quality of education.............	1	2	3	4	20/
i. Traffic congestion...............	1	2	3	4	21/
j. Cost of housing..................	1	2	3	4	22/

We are interested in your attitudes toward pollution. By pollution I mean air pollution, water pollution, noise pollution (which is annoying sounds such as traffic noise or jackhammers), and street garbage pollution. By street garbage pollution I mean such things as cans, paper, bottles, litter and construction junk which you find around outside your home or on the street.

2. How serious a problem is pollution is your area? Would you say it was...
REPEAT FOR EACH CATEGORY BELOW

	Very serious	Somewhat serious	Or, Not serious at all	DK/NA	
a. Pollution.......................	1	2	3	4	23/
b. Water pollution..................	1	2	3	4	24/
c. Air pollution....................	1	2	3	4	25/
d. Noise pollution..................	1	2	3	4	26/
e. Street garbage pollution.........	1	2	3	4	27/

3. Do you think that your neighbors and friends worry or think about pollution more than you, less than you, or about the same as you?

More than you................. 1	28/
Less than you................. 2	
About the same as you......... 3	
DK/NA......................... 4	

4. In which of the following ways does pollution bother you? Does it:

	Yes	No	DK	
a. Make you mad?........	1	2	3	29/
b. Make you annoyed?...	1	2	3	30/
c. Make you depressed?.	1	2	3	31/

5. What kind of action would you consider taking against pollution in your area? Would you: READ AND RECORD EACH CATEGORY. IF RESPONSE IS 'YES' ASK IMMEDIATELY How effective do you think it would be? Would you say it would be very effective, somewhat effective, or not very effective at all? CIRCLE IN COLUMNS 5-7,9

				IF YES ANSWERS				
	Yes	No	DK	VE	SE	NVE	DK	
a. Complain to the authorities?...	1	2	4	5	6	7	9	32/
b. Stay indoors more frequently?...	1	2	4	5	6	7	9	33/
c. Avoid swimming in Lake Michigan?	1	2	4	5	6	7	9	34/
d. Sign petitions?................	1	2	4	5	6	7	9	35/
e. Use a car less frequently?.....	1	2	4	5	6	7	9	36/
f. Pay high prices for goods so that industry could recycle and install anti-pollution equipment......................	1	2	4	5	6	7	9	37/
g. Join a neighborhood or environmental group who is concerned about pollution problems in your area?.....................	1	2	4	5	6	7	9	38/
h. Avoid boating in Lake Michigan?	1	2	4	5	6	7	9	39/
i. Keep your windows closed more often?.........................	1	2	4	5	6	7	9	40/
j. Attend local government meetings and hearings?................	1	2	4	5	6	7	9	41/
k. Get out of the city more often?	1	2	4	5	6	7	9	42/
l. Pay more taxes for clean-up programs?......................	1	2	4	5	6	7	9	43/
m. Reduce the amount or electricity you use?.....................	1	2	4	5	6	7	9	44/
n. Vote for environmental legislation and/or people which would result in a cleaner environment?....	1	2	4	5	6	7	9	45/

6. Have you ever done any of the actions mentioned above?

Yes.(ASK A)...... 1	46/
No..(SKIP TO Q.7). 2	

IF YES, ASK
A. What have you done? RECORD VERBATIM

____ ____ 47-48/

____ ____ 49-50/

____ ____ 51-52/

7. People give many reasons for not doing more against pollution. Which of the following reasons best explain why you haven't done more? Would you say it was because there was too much hassle and/or red tape, not enough time, wouldn't do any good, or because you are really not interested? CIRCLE AS MANY AS APPLY

Too much hassle (red tape).......... 1	53/
Not enough time...................... 2	
Wouldn't do any good................. 3	
Not really interested................ 4	
DK/NA................................ 9	

8. The following questions are related to your attitude toward street garbage pollution. By street garbage pollution I mean such things as cans, paper, bottles, litter and construction junk which you find around outside your home or on the street. What kind of action would you consider taking against street garbage pollution? Would you: READ AND RECORD EACH CATEGORY. IF RESPONSE IS 'YES' ASK IMMEDIATELY How effective do you think it would be? Would you say it would be very effective, somewhat effective, or not very effective at all? CIRCLE IN COLUMNS 5-7,9.

	Yes	No	DK	IF 'YES' ANSWERS				
				VE	SE	NVE	DK	
a. Pay a refundable deposit on all beverage containers?...........	1	2	4	5	6	7	9	54/
b. Vote for legislation regarding street garbage?................	1	2	4	5	6	7	9	55/
c. Complain to authorities about litter?.......................	1	2	4	5	6	7	9	56/
d. Separate your trash so that cans jars, bottles and newspapers could be recycled?.............	1	2	4	5	6	7	9	57/
e. Stop using disposable products such as paper cups, paper plates and plastic utensils?..........	1	2	4	5	6	7	9	58/
f. Pick up litter on the street?..	1	2	4	5	6	7	9	59/
g. Pay to have your trash separated so that it could be recycled?..	1	2	4	5	6	7	9	60/
h. Ask someone to place more trash containers on the street?	1	2	4	5	6	7	9	61/
i. Require people who walk their dogs to clean up after them?...	1	2	4	5	6	7	9	62/
j. Pay higher prices for goods which would last longer?.......	1	2	4	5	6	7	9	63/

9. Have you ever done any of the actions mentioned above?

Yes (ASK A)........1	64/
No (SKIP TO Q.10).2	

IF YES, ASK
A. What have you done? RECORD VERBATIM

_____ _____ 65-66/

_____ _____ 67-68/

_____ _____ 69-70/

10. People give many reasons for not doing more against street garbage pollution. Which two of the following reasons best explain why you haven't done more? Would you say it was because of too much hassle and/or 'red tape', not enough time, it wouldn't do any good, or because you are not really interested? CIRCLE ALL THAT APPLY AND PROBE

	Yes	No	DK/NA	
Too much hassle	1	2	9	71/
Not enough time......	1	2	9	72/
Wouldn't do any good.	1	2	9	73/
Not really interested	1	2	9	74/

11. Are there any other reasons? RECORD VERBATIM

_____ 75/

_____ 76/

12. Would you be willing to respond in greater detail regarding your opinion of pollution sometime in the near future?

Yes. (ASK A)......... 1   77/
No................... 2

IF YES, ASK

What is your full name and address?

Name_____

Address_____

THANK YOU

INTERVIEWER COMMENTS:

INTERVIEWER NAME:_____

Approximate Time of Interview _____ minutes   Date of Interview May ____

APPENDIX 3

THE CHICAGO STUDY AREAS

A typology of eight types based on pollution levels, social status, and race was utilized. The twenty-two community areas (fig. 2) finally selected exemplify each of the eight types, which are as follows:

Type 1. High pollution, low socioeconomic status, white
Example: Uptown

Type 2. High pollution, low socioeconomic status, black
Examples: West Garfield Park, East Garfield Park, Lawndale, Woodlawn, and Englewood

Type 3. Low pollution, low socioeconomic status, white
Examples: Burnside and Pullman

Type 4. Low pollution, low socioeconomic status, black
Example: Avalon Park

Type 5. High pollution, high socioeconomic status, white
Examples: Lincoln Park, Near North, and Hyde Park

Type 6. High pollution, high socioeconomic status, black
Examples: Kenwood and South Shore

Type 7. Low pollution, high socioeconomic status, white
Examples: Forest Glen, Dunning, Montclare, Garfield Ridge, Clearing, and Beverly

Type 8. Low pollution, high socioeconomic status, black
Examples: Roseland and Calumet Heights

Owing to the diversity in some of these community areas, these types do not reflect the entire composition of the community areas. Instead, we used those census tracts that were most consistent with the category in which the area was placed. For example, Hyde Park is a racially mixed community, with the whites

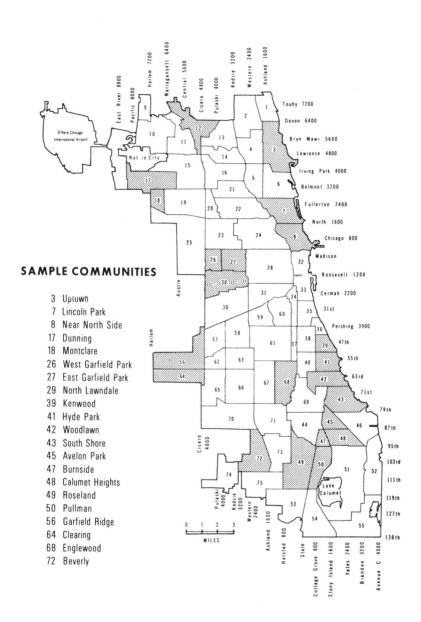

Fig. 2. Community areas in Chicago

enjoying a high status. Therefore the area south of 55th Street, from the lake on the east to Ellis Avenue on the west, was selected for the survey.

Now let us turn to a brief description of the community areas selected for inclusion in this study.

Type 1: High Pollution, Low Socioeconomic Status, White

Uptown

Uptown (community area 3) is a study in contrasts. The northern part of the community has an average yearly income of $15,000, while the southern section averages only $8,000. The specific census tracts included in this study are bounded by Irving Park on the south, Lawrence Avenue on the north, Halstead Street on the east, and Clark on the west. The area is almost all white, with a large percentage of whites who have migrated from the South. There is also a sizable Latin American and Asian ethnic population. The largest age group consists of adults twenty-five to forty-four, and there is a substantial school-aged population. Most of the housing units are renter-occupied, with a median rent of $110. Less than 5 percent of the population own their own homes. The area is highly transient; significant number of inhabitants have moved in since 1970. Most of the residents have high-school educations and are employed in clerical and technical occupations. Incomes average between $8,500 and $9,500 a year. In the parts of the area specifically selected for this study, between 20 and 30 percent of the families have an income below the poverty level, and between 10-15 percent are unemployed.

Type 2: High Pollution, Low Socioeconomic Status, Black

West Garfield Park

West Garfield Park (community area 26) is due west of

Chicago's Loop. The community is entirely black (more than 90%), with an average yearly income of less than $7,500. Between 25 and 30 percent of the families in the area have incomes below the poverty level. Unemployment averages between 20 and 30 percent. The population falls into two main age groups, those between twenty-five and forty-four, and children between five and fourteen. Most of the people live in apartments, with very little crowding, and they pay an average rent of $100. Most of the influx into the area occurred in 1960-64 and 1969-70. A majority of the migrants were blacks from the South. The median school years completed is less than 9.5. Those of the population who are employed are operatives or clerical workers in the wholesale and manufacturing industries. In the decade between 1960 and 1970, the area completely changed from all white to all black. The census tracts under study are the area bounded by Madison on the south, Kinzie on the north, Pulaski on the west, and Hamlin on the east and the area bounded by Hamson on the north, Taylor on the south, Independence on the east, and Pulaski on the west.

East Garfield Park

East Garfield Park (community area 27) is also entirely black, with family incomes ranging between $5,700 and $7,200; 30 to 40 percent of the population have incomes below the poverty level. As in West Garfield Park, the population falls into two main age brackets--those between twenty-five and forty-four and those between five and fourteen, the school-aged poulation. Most of the people in the area rent apartments and pay an average of $100 a month. The median number of school years completed is fewer than ten. Unemployment ranges from 20 to 30 percent. Those who are employed are in the wholesale trade and manufacturing industries. The largest influx of people into the area occurred from 1960 to 1964 and from 1969 to 1970. Most of

the incoming blacks were either from Chicago or from the South.
The area under study is bounded by the Chicago and North Western
railroad tracks on the north, Rockwell on the east, Madison on
the south, and Kedzie on the west. The other area of study is
bounded by the Chicago and North Western railroad tracks on the
east, Jackson on the north, Taylor on the south, and Kedzie on
the west. The community underwent racial change in the decades
from 1950 to 1970. In 1960 two-thirds of the community was black,
and by 1970 the entire community was black.

North Lawndale

North Lawndale (community area 29) is situated west of
Chicago's Loop. The population is predominantly black (more
than 90%) and poor (average incomes are less than $7,500); 30
to 35 percent of the population is below the poverty level. As
with East and West Garfield Park, the population in North Lawndale
falls into two main age categories: twenty-five to forty-four,
and children five to fourteen. Most of the people rent apartments
in the area and pay between $90 and $110 per month.

The majority of the black population moved into the area
in 1960-66 and 1969-70, although in 1960 the area was already
91% black. It was not until 1970 that the area was totally
converted. The average years of school completed is ten.
Unemployment in North Lawndale runs from 25 to 35 percent. Those
employed are mostly blue-collar workers in the manufacturing and
wholesale trade industries. The specific census tracts selected
for study are bounded by Arthington on the north, 16th Street
on the south, Pulaski on the west, and California on the east.

Woodlawn

Woodlawn (community area 42) is on Chicago's lakefront,
south of the Loop. More than 85 percent of the population is
black and poor (incomes less than $6,500). In the area selected

for study (bounded by 60th Street on the north, 67th Street on the south, Lake Michigan on the east, and Ellis Avenue on the west), 30 percent of the families have incomes below the poverty level. Most residents live in apartments and pay from $100 to $140 rent per month. The residents fall into three main age categories: school-age children five to fourteen, adults twenty-five to forty-four, and the aging population forty-five to sixty-four. The median schooling completed is eleventh grade. Unemployment in the area is high, about 25 percent. Those people who do work are mostly in clerical positions (e.g. clerks and typists at the University of Chicago) and in the wholesale trade industries. There has been massive demolition along 63rd Street and in other parts of the community owing to urban renewal efforts beginning in early 1970. As of 1976 this land is still largely vacant.

Englewood

Englewood (community area 68) is characterized by a poor (incomes under $8,000) black population. Between one-fifth and one-fourth of the families in the area have incomes below the poverty level. Residents pay an average of $110 per month in rent for their apartments. The age distribution falls into two main categories, as in the other communities of this type: ages twenty-five to forty-four and five to fourteen. The median schooling completed is tenth grade. Unemployment averages about 20 percent, and those who are employed work in the manufacturing and wholesale and retail trade industries. Blacks began migrating into the area in the 1950s, and there were substantial in-migrations in 1960-66 and in 1969-70. The specific area under study is bounded by 55th Street on the north, 71st Street on the south, Halsted Street on the west, and Yale Avenue on the east.

Type 3: Low Pollution, Low Socioeconomic Status, White

Burnside

Burnside (community area 47) is a low-income, working-class area on Chicago's South Side. The residents are mostly white persons of Hungarian, Polish, Russian, or Italian descent. Almost half own their own homes, which are valued at an average of $15,700. The median rent is $90. Median years of schooling completed is 9.4. Unemployment is low, generally below 5 percent. Workers in the area are employed in the manufacturing industry, primarily in the steel works on the city's South Side. The age structure of the community is divided into two main groups: parents aged twenty-five to forty-four and children between five and fourteen. The area primarily consists of single-family residences built in the 1950s, when migration into the area was at its peak. The area selected for study includes the entire community, bounded by 87th Street on the north, 95th Street on the south, Illinois Central Gulf railroad tracks on the west, and the New York Central and Saint Louis railroad tracks on the east.

Pullman

Pullman (community area 50) is undergoing drastic change. The area has three census tracts: the northernmost tract is entirely black, the middle tract is one-fourth black, and the southern tract is entirely white. The area included in this study is the southern portion of Pullman, bounded by 111th Street on the north, 115th Street on the south, Cottage Grove on the west, and Doty (Calumet Expressway) on the east. Pullman is a white low-income, working-class community. Incomes are less than $11,000, and unemployment is generally low (less than 7%). Twenty-six percent of the residents own their own homes, which are valued at an average of $11,800, while most of the residents rent apartments. Rents average about $90. There are three main age groups in the area: parents between the ages of twenty-five and forty-four, the aging population forty-five to sixty-four, and children five to fourteen. Ethnically, Pullman consists of white Americans of Southern origin, Italians, and some Spanish-Americans. Most residents are employed in the manufacturing industries, primarily the Sherwin-Williams paint factory, which is in the southern portion of the area.

Type 4: Low Pollution, Low Socioeconomic Status, Black

Avalon Park

Avalon Park (community area 45), on the city's South Side, is predominantly a black working-class neighborhood. Mean incomes are in the $10,000 range, with only 7 percent of the families in the area having incomes below the poverty level. The area chosen for study is a lower-middle-class area bounded by 76th Street on the north, 87th Street on the south, Stony Island Avenue on the east, and the New York Central and Saint Louis Railroad tracks on the west. In this area half the residents own their own single-family homes, valued at an average of $18,000. Apartment rents average $120 per month. The age structure of the community is largely in two groups: parents aged twenty-five to forty-four and children five to fourteen. Most of the residents have completed high school. The residents are employed in white-collar occupations, mainly clerical and technical. The area changed from predominantly white collar predominantly black in the decades following 1960, and by 1970 more than 75 percent of the population was black.

Type 5: High Pollution, High Socioeconomic Status, White

Lincoln Park

Lincoln Park (community area 7) is situated north of the Loop along Lake Michigan. It is a predominantly white upper-class area of single professional people. Average incomes are about $20,000. The median years of schooling completed is fourteen. Most of the residents live in apartments and pay an average of $160 per month in rent. Like some of the communities discussed earlier, Lincoln Park is undergoing transition. For this reason, those characteristics of the community that were listed apply only to that area nearest the lake and Lincoln Park, bounded by North Avenue on the south, Diversey on the north, Halsted on the west, and Lincoln Park on the east. The residents are white-collar clerical workers or professionals. The area as a whole has experienced a decline in population, but this has happened west of the area under study. The age of the population is mainly between twenty-five and sixty-four, and there are very few children.

Urban decay and expansion of the black population eastward from the Chicago River has led to urban renewal projects. Today there is a sizable Latin and black community in the Lincoln Park area, although they live west of Clark Street. The affluent whites maintain their hold along the lakefront.

Near North

The Near North (community area 8) is just north of the Loop. The particular area under study, bounded by North Avenue on the north, the Chicago River on the south, Lake Michigan on the east, and State Street on the west, is better known as the "Gold Coast." As a whole, the Near North is a diverse community. It includes both the high-income white area known as the Gold Coast and the low-income black housing project, Cabrini Green Homes, less than one mile away. For this type, we will look expressly at the Gold Coast area.

The area is entirely white upper class. Incomes range from $30,000 upward. Most of the residents live in apartments or condominiums with values upward of $50,000. The main age structure is from twenty-five to sixty-four with very few children present. Most of the residents have at least two years of college. This is a white-collar professional area whose residents are employed in the Loop and use public transportation to get there.

Since 1930 there has been substantial commercial development along North Michigan Avenue, catering to the Gold Coast population. The latest construction was Water Tower Place, an exclusive shopping area adjacent to the Ritz-Carlton Hotel. Also, the Northwestern University Chicago campus has expanded and added a medical center.

Hyde Park

Hyde Park (community area 41) is similar to the other two areas in this type in that it has undergone racial change. Hyde Park is a racially mixed community, with blacks dominating the northern portion and whites the southern portion. For this study we chose the southern portion of the community, bounded by 55th Street on the north, 59th Street on the south, the lake on the east, and Ellis Avenue on the west.

Incomes in this area average $20,000. Most residents live in apartments and pay an average rent of $170 per month. Because the University of Chicago is in the area, the population tends to reflect a student transience. The year-round residents tend to live in single-family homes or condominiums valued at an average of $46,000. Most residents have completed college, and some have done graduate work. The age structure of the community is divided into two main components: those twenty-five to forty-four and those forty-five to sixty-four. Most residents are employed in professional or clerical jobs and either walk to their place of employment, mainly the University of Chicago, or take public transportation to the Loop. There is a substantial

Jewish resident population.

Type 6: High Pollution, High Socioeconomic Status, Black

Kenwood

Kenwood (community area 39) is situated along the southern shore of Lake Michigan. It has a wide diversity in its social composition, and for that reason one small portion of the community, bounded by 47th Street on the north, 51st Street on the south, Woodlawn on the east, and Ellis on the west, was chosen as an example of this type.

Incomes in this area average $17,000. The population is more than 75 percent black. Single-family homes are large estates valued at more than $50,000, but only 20 percent of the population own their own homes. Renters generally pay from $130 to $160 per month. The population in the area has declined steadily, although the area is relatively stable. The residents are white-collar workers employed in clerical or professional positions (primarily at the University of Chicago).

South Shore

South Shore (community area 43) is also situated on the city's South Side. Like some of the other communities under study, South Shore is racially mixed. To make it fit this type, two specific areas were chosen for examination: one is bounded by 67th Street on the north, 71st Street on the south, Stony Island Avenue on the west, and the lake on the east; the other is bounded by 73rd Street on the north, 79th Street on the south, Jeffrey on the west, and Yates on the east.

In these areas blacks make up about 60 percent of the total population and have mean incomes of $14,000. Homes in the area are valued at more than $25,000, although only a small portion of the residents own their own homes. Apartment-dwellers generally pay $150 a month in rent. The age structure of the community is mainly adult, with ages from twenty-five to sixty-four. Most of the residents have moved into the community since 1969 and have come from other parts of the city. Most of the workers are clerical workers employed in professional industries, primarily in Hyde Park (the University of Chicago) and in the Loop.

Type 7: Low Pollution, High Socioeconomic Status, White

Forest Glen

Forest Glen (community area 12) is an all-white community on the city's northwest side. Incomes average $20,000. More than 80 percent of the residents own their own homes, valued at an average of $37,000. The residents are mainly of German, Polish, or Italian ancestry. Residents fall into the aging category, with most between forty-five and sixty-four years of age. There are very few children in the area. Forest Glen is a stable community, most of the residents having moved into the area from other parts of the city in the 1950s. Most residents have completed high school, and most are white-collar workers in clerical and professional positions in the wholesale and retail trade industries. The area sampled for this study is north of the North Branch of the Chicago River and west of Cicero Avenue.

Dunning and Montclare

Dunning (community area 17) and Montclare (community are 18) are neighboring areas so similar in characteristics that they will be considered together in this brief analysis.

Both areas are totally white, and incomes are in the $13,500 range. More than 60 percent of the residents own their own homes, which have an average value of $30,000. As in Forest Glen, most residents fall in the forty-five to sixty-four age bracket, although there is a sizable population aged twenty-five to forty-four, with some children. Poles are the dominant group, followed by Germans and Italians. The nieghborhood is still growing and is relatively stable. Most of the inhabitants moved into the area in the 1950s. Most of the residents have high-school educations and are employed as clerks, craftsmen, and foremen in the manufacturing and wholesale and retail trade industries.

The area designated for study in Dunning was bounded by Austin Avenue on the east, Harlem Avenue on the west, Addison on the north, and Belmont on the south. In Montclare the area of study was the entire community except for the area between Diversey and Fullerton.

Garfield Ridge

Garfield Ridge (community area 56) is a white Polish and Italian community on the city's southwest side. The residents earn an average income of $14,300. More than three-fourths own their own homes, which have an average value of more than $23,000. The population is generally older, forty-five to sixty-four, although there are families with school-age children. Most residents have completed high school and are employed as clerical help in manufacturing industries. The area has been stable for many years; most of the inhabitants have moved into the area since 1950.

The area sampled is divided into three portions: one bounded by 47th Street on the north, 51st Street on the south, Cicero on the west, and the Beltline railroad on the east; another bounded by 51st Street on the north, 55th Street on the south, Cicero on the east, and Laramie on the west; and the third bounded by 55th Street on the north, 59th Street on the south, Narragansett on the west, and Central on the east.

Clearing

Clearing (community area 64), like its neighbor to the north, Garfield Ridge, is an all-white Polish and Italian community. The average income in the area is $13,600. Seventy-five percent of the residents own their homes, which have an average value of $23,000. The residents are primarily in the forty-five to sixty-four age group, but there are younger families with children. Like Garfield Ridge, Clearing is a stable community, although it is newer, most residents having moved into the area between 1960 and 1964. Most of the adults have finished high school and are employed as clerical workers in manufacturing and trade industries. The specific area chosen for study is bounded by 59th Street on the north, 63rd Street on the south, Harlem Avenue on the west, and Central Avenue on the east.

Beverly

Beverly (community area 72) is an all-white community on the city's southwest side. Most of the inhabitants belong to no particular ethnic group (they are not first- or second-generation ethnics), except for the Irish. More than 90 percent of the residents own their own single-family homes valued at an average of $23,000. The residents earn an average income of $20,000. The population is generally older (forty-five to sixty-four), but there are younger families with children. Most of the adults have completed high school or some college. The residents are employed in clerical and technical occupations in the trade and professional industries. Beverly is a racially stable community, although there has been some black encroachment from the east. Most of the residents moved into the area during the 1950s. The subarea selected for study is bounded by 87th Street on the north, 107th on the south, Western on the west, and Damen on on the east.

Type 8: Low Pollution, High Socioeconomic Status, Black

Calumet Heights

Calumet Heights (community area 48) is a black (75%) working-class area on the city's South Side. Incomes average about $15,000. About 40 percent of the residents own their homes, valued at an average of $20,000, and the rest are apartment-dwellers who pay an average of $145 a month in rent. Three age groups dominate the community: an aging population aged forty-five to sixty-four and younger families with parents between twenty-five and forty-four and children aged five to fourteen. The adult population is well educated, having completed high school and some college. Most of the inhabitants moved to Calumet Heights from other parts of the city in the late 1960s. Most of the work force in the community consists of clerical workers employed in professional and wholesale and retail trade industries. The area went from completely white to partially black in the decade following 1960. The eastern section of the community is a Spanish community, and the area selected for study (that bounded by 87th Street on the north, 95th Street on the south, Stony Island on the east, and South Chicago Avenue on the west) is entirely black.

Roseland

Roseland (community area 49) is on the city's South Side and is primarily black, although the southernmost census tracts in the community are entirely white. The area included in this study is the black area from 95th Street on the north to 99th Street on the south, Stewart Avenue on the west, and Cottage Grove on the east. In this area incomes average $14,300 per year, and more than 75 percent of the residents own their own homes, valued at an average of $23,000. The age structure again falls into three main groups: those between forty-five and sixty-four, those between twenty-five and forty-four, and those five to fourteen. Most of the residents moved into the area from 1950 to 1959 and again from 1969 to 1970, coming from other parts of the city. Racial change occurred in the decade following 1960. Most of the adults have completed high school and are employed as clerical workers in the professional and wholesale and retail trade industries.

APPENDIX 4

THE MEASURES OF ENVIRONMENTAL QUALITY

Four types of pollution were measured: air, water, noise, and solid waste. Also available were indicators of rat bite and pesticide poisoning risk and levels of hazardous chemicals in the air. The information comes from a research project--in which I took part--conducted from 1974 to 1976. The complete results of the project can be found in The Social Burdens of Environmental Pollution, edited by Brian J. L. Berry (1977), where the methodology, sources, analysis, and interpretation of the data are discussed. This section serves as a summary of the results for Chicago. More current data were used for air and water quality.

Air Quality

Nine measures of air quality were used, including particulates (1974 and 1975), carbon monoxide (1974 and 1975), nitrogen dioxide (1974 and 1975), sulfur dioxide (1974 and 1975), and ozone (1975). Using these parameters, two indexes of air quality were constructed. The first index is a compilation of the number of violations of federal standards for each of the nine parameters. The second index uses only the five parameters monitored for 1975.

In both of the control years the federal standards for sulfur dioxide (0.03 ppm arithmetic mean) were met for the entire city. For particulates there is a different story. A comparison of the distribution of particulate pollution for the two years points out the areas with the highest concentrations.

For 1974 the highest concentrations are in the Loop and the Near North area and in the industrial sector of the city along Lake Calumet and the Indiana border (fig. 3). The 1975 data reveal that whereas the Loop now violates only secondary standards, the southeast industrial area shows little improvement. Most of the central and southern sections of the city still violate the federal primary (76-100 $\mu g/m^3$) and secondary (>100 $\mu g/m^3$) standards (fig. 4).

Carbon monoxide was found to be a continuing problem (exceeding the standard of 9 ppm per eight-hour average in 1974 and 1975) in the Loop and those areas that border the main freeways in the city; the Edens and Kennedy to the north, the Stevenson and Eisenhower to the west, and the Dan Ryan to the south and southwest. The Near North was the only community in this study that violated the nitrogen standard (0.05 ppm arithmetic mean) for both 1974 and 1975. All other communities met the standard for those years. Finally, ozone was found to be a problem in the community areas to the west and to the south from the Loop along the lake. The Kenwood High School monitoring station recorded the highest levels of ozone. This directly and indirectly (owing to prevailing wind patterns) affects Kenwood, Hyde Park, Woodlawn, and South Shore.

The cumulative air pollution index clearly shows the areas with the worst air quality: Forest Glen, Near North, East Garfield Park, North Lawndale, and Beverly (fig. 5). By far the worst air quality in the city is found on the Near North, where six of the nine parameters exceeded federal standards, primarily because of its proximity to the Loop and the daily influx of commuters using automobiles. As mentioned before, Forest Glen and Beverly are bounded by major freeways, and thus reflect the high concentrations of carbon monoxide and other pollutants emitted by automobiles. The remaining two communities

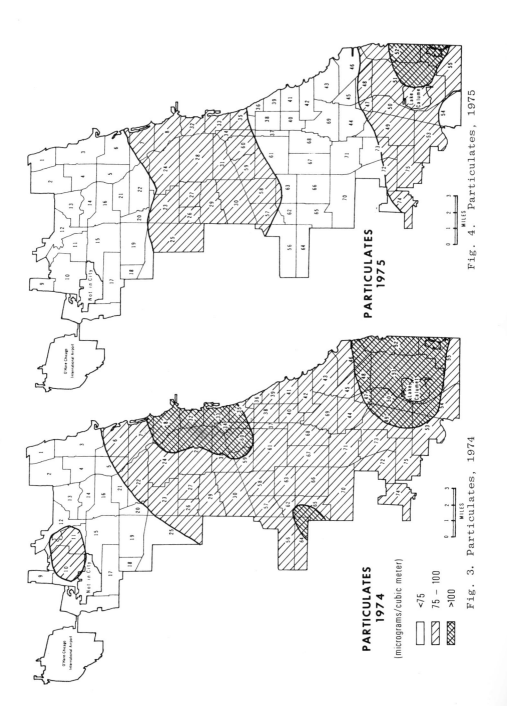

Fig. 3. Particulates, 1974

Fig. 4. Particulates, 1975

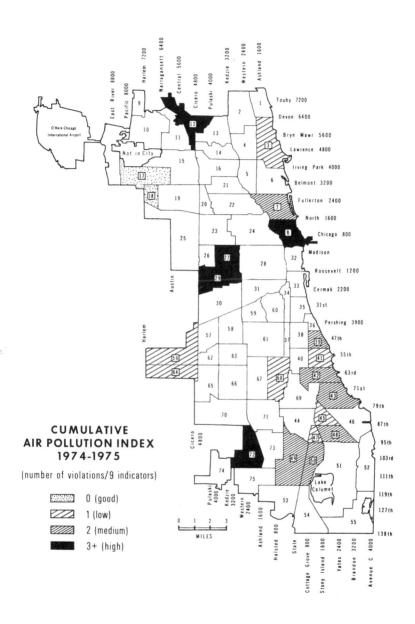

Fig. 5. Cumulative air pollution index

near major transportation arteries, East Garfield Park and North Lawndale, also have very high ozone levels. The three areas in the vicinity of Lake Calumet (Calumet Heights, Roseland, and Pullman) show high levels of particulate pollution owing to the large amount of heavy industry in that area.

When we compare the 1975 index to the cumulative index we find basically the same pattern, although Forest Glen and Beverly are no longer in the high-pollution category (fig. 6). Again, the Near North has the most violations (three, followed by East Garfield Park and North Lawndale, each with two.

Trace Metals in the Atmosphere

Information on trace metals in the atmosphere was available for 1971 only. Although the data are not up to date, no more current information is available. An index of exposure to trace metals was compiled for mercury, arsenic, zinc, manganese, lead, cadmium, copper, and iron.

The areas of highest exposure are on the South Side in the immediate vicinity of Lake Calumet, a highly industrialized area (fig. 7). Arsenic and cadmium levels are very high in Uptown because of industrial activity and the combustion of coal in power plants. Dunning, one of the other three areas of medium exposure on the north side of the city, recorded high levels of copper and mercury from industrial applications, lead from automobile exhaust and incineration of refuse, and zinc from industrial applications. Lead and zinc were also predominant in West and East Garfield Park. The communities of Beverly, Roseland, and South Sore on the South Side were in the medium-exposure category owing to heavy concentrations of mercury, copper, iron, and manganese from industry in the Lake Calumet region.

The three communities bordering the industrial area of

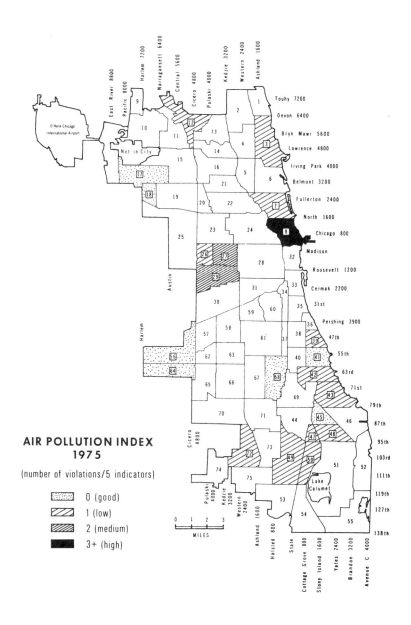

Fig. 6. Air pollution index for 1975

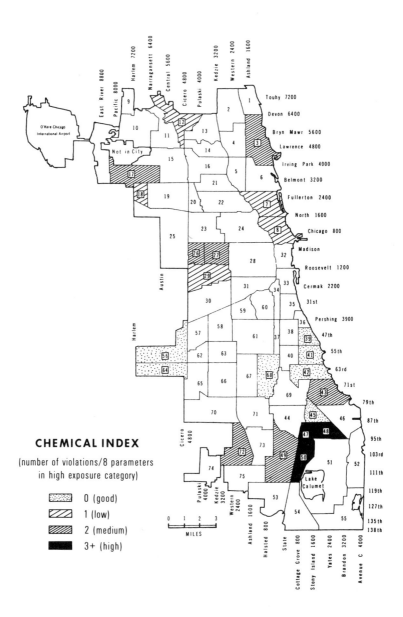

Fig. 7. Index of trace metals in the atmosphere

Lake Calumet recorded the highest levels of mercury, copper, manganese, and iron in the entire city. Of all the communities studied, Pullman had the highest exposure to trace metals in the atmosphere, followed by Calumet Heights and Burnside.

Water Quality

Though it is easy to measure water quality and monitor it for possible contaminants, it is not so easy to determine how much humans are exposed to water pollution. Previous authors have used water use (David 1971) and distance from a body of water (Jacoby 1972) as measures of exposure levels. In this study we have selected the latter approach. We compiled twenty-four parameters for water quality into one index, then adjusted it for nearness to the body of water. Also incorporated in the analysis were two distance measures for each community: distance from Lake Michigan and distance from the nearest body of water.

The twenty-four parameters used were divided into seven major categories: oxygen measuring; microbiological; acidity/alkalinity; solids; organic chemicals (cyanide, methylene blue active substances, oil and grease, and phenols); dissolved elements, ions, and compounds (chloride, fluoride, ammonia, nitrate, phosphorus, and sulfate); and trace metals (arsenic, cadmium, copper, iron, lead, manganese, mercury, nickel, silver, and zinc). The water-pollution index was constructed by determining whether the parameter met or exceeded federal standards; the number failing the standards was summed over the twenty-four parameters. For a more complete and detailed analysis of water quality in the Chicago area as well as background discussion of the parameters utilized in this present research, consult <u>The Social Burdens of Environmental Pollution</u> (Berry 1977), specifically chapter 4.

The water pollution index shows two areas of high water

pollution: South Shore and Calumet Heights (fig. 8). The Near North has a severe water pollution problem originating from the North Branch of the Chicago River, not from Lake Michigan. Since the census tracts studied for this community are along the lakefront, the severe water pollution problem does not show up in the questionnaire response, although it does deserve some mention. The bulk of the pollution of the river comes from waste material from treatment plants, detergents, and industries along the river that discharge heavy metals into the water.

South Shore's water pollution problem is with Lake Michigan. This particular portion of the lake is high in chloride, ammonia, sulfate, phosphorus, and trace metals. The other community with a high pollution index rating, Calumet Heights, is situated one to three miles from the Calumet River; but the river exceeds the standards for so many parameters that the proximity measure is dwarfed. High on the list of exceeded standards are cyanide and phenols, oil and grease, and chlorides and fluorides. Those communities that are more than three miles from the water generally have no water pollution problem, whereas those in close proximity are most likely to face this problem, the extent naturally depends on the number of violations of the standards.

Noise Pollution

Data on noise pollution in Chicago are rather limited, although detailed information is available for part of the South Side, part of the West Side, and one part of the North Side. Again, the reader is encouraged to consult The Social Burdens of Environmental Pollution (Berry 1977), particularly chapter 6. Five measures were used in determining noise pollution: daytime background noise levels; background noise levels exceeding the standards; intermittent noise levels; complaints about noise;

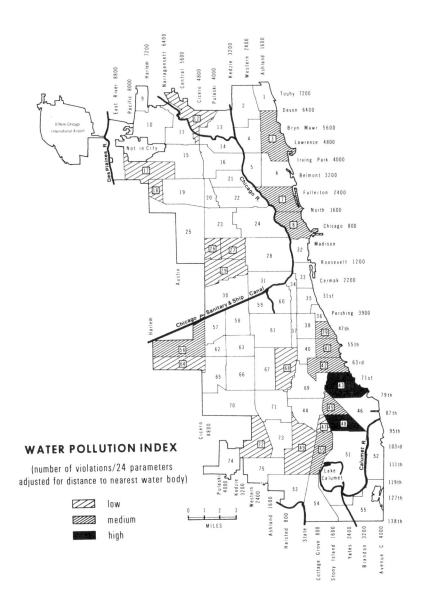

Fig. 8. Water pollution index

and airport noise exposure. The noise index used in this study relied on only one of these measures: daytime background noise; there simply was not any information on the other measures for some of the communities.

Ten of the communities appear to have noise pollution problems that are serious (East and West Garfield Park) to semi-serious (Forest Glen, Dunning, North Lawndale, Woodlawn, Avalon Park, Garfield Ridge, Clearing, and Englewood) (fig. 9), primarily because they have residential areas close to industrial and commercial establishments or heavily used thoroughfares. East and West Garfield Park have very high (over 62 dBA) daytime background noise levels. Intermittent noise levels exceed 80 dBA in four communities: Avalon Park, Burnside, Calumet Heights, and Pullman. This is primarily because of the heavy industry in these communities. Complaints about noise came from the west and north sides of the city, although no spatial pattern was apparent. The community of Forest Glen is exposed to aircraft noise--it falls within the 30 NEF contour for O'Hare Airport. Two southwestern communities, Clearing and Garfield Ridge, have some noise problem from Midway Airport, but since it is no longer used by commercial airliners no information was available on the noise exposure forecasts.

Solid Waste Pollution

Data on solid waste pollution cover the years 1972 and 1973. The last systematic data were gathering in 1972 in conjunction with an air pollution study. The index of solid waste pollution was compiled using indicators of total residential solid waste; total commercial solid waste; total combination solid waste; total industrial solid waste; and total subsidiary solid waste (e.g., bulk trash, abandoned cars, street dirt, tree and stump removal, and residential demolitions). This index

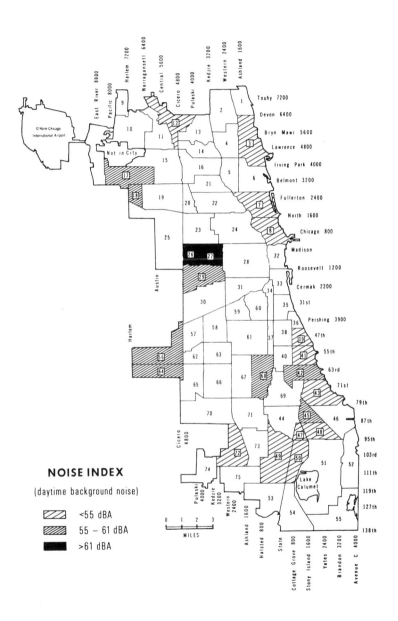

Fig. 9. Noise pollution index

basically shows the total amount of solid waste in the city.

High levels of residential solid waste or high exposure to it were found in all the communities that border Lake Michigan and in West Garfield Park. Since generation primarily depends on density, this finding was not surprising, as the highest densities of dwelling units are along the lakefront. Commercial solid waste was found to be a problem in five communities--Near North, East and West Garfield Park, Hyde Park, and Englewood--communities with a large proportion of commercial establishments such as theaters and stores. Combination solid waste (apartments over commercial establishments in the same building) was found to be a problem in only two areas, Uptown and West Garfield Park. Industrial solid waste was a major component of the total solid waste for East Garfield, North Lawndale, and Pullman, communities with a large percentage of industry.

When one looks at subsidiary sources of solid waste a different pattern emerges. The highest totals of subsidiary solid waste are found in Kenwood (undergoing urban renewal) and Englewood (a black area suffering from urban decay). The next highest levels are found in Uptown, East and West Garfield Park, and Hyde Park. Englewood is in the highest category in all the components except tree and stump removal. Bulk trash appears to be a major problem in Englewood and much less of a problem in Kenwood, although both areas record very high values. Abandoned cars create a solid waste problem in Uptown, Lincoln Park, East Garfield Park, and Englewood. Tree and stump removal is a major source of solid waste in Woodlawn and Avalon Park (primarily owing to diseased trees and urban renewal). And residential demolitions contribute significantly to solid waste pollution in Kenwood, East Garfield Park, and Englewood.

When looking at the spatial incidence of solid waste pollution one can readily see the emergent pattern (fig. 10).

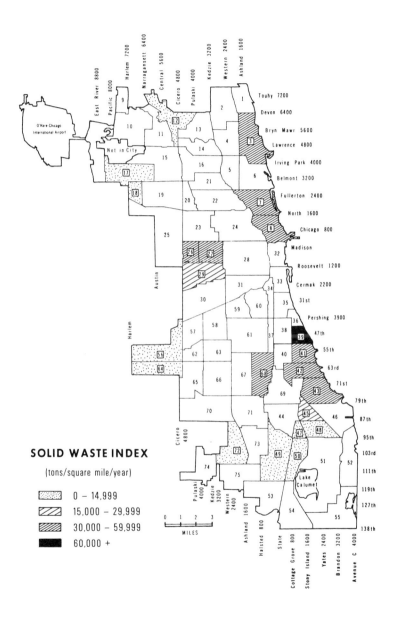

Fig. 10. Solid waste pollution index

Solid waste is a major problem in inner-city neighborhoods where densities are very high (Lincoln Park and Near North), where there is urban redevelopment owing to changes in neighborhood composition and urban decay (East and West Garfield Park, Englewood, and Kenwood), or a combination of the two (Woodlawn and South Shore).

The Social Burdens of Environmental Pollution (Berry 1977), particularly chapter 7, gives more detail about the solid waste problem in Chicago as well as a discussion of the spatial patterns as related to socioeconomic factors.

SELECTED BIBLIOGRAPHY

SELECTED BIBLIOGRAPHY

Abelson, R. 1959. Modes of resolution of belief dilemmas. Journal of Conflict Resolution 3: 343-52.

Abelson, R., and Rosenberg, M. J. 1958. Symbolic psycho-logic: A model of attitudinal cognition. Behavioral Science 3: 1-13.

Ajzen, I., and Fishbein, M. 1970. The prediction of behavior from attitudinal and normative variables. Journal of Experimental Social Psychology 6: 466-487.

Althoff, Phillip, and Greig, William H. 1974. Environmental pollution control policy making: An analysis of elite perceptions and preferences. Environment and Behavior 6, 3 (September): 259-88.

Ashmore, Richard D., and McConahay, John B. 1975. Psychology and America's urban dilemmas. New York: McGraw-Hill.

Auliciems, Andris, and Burton, Ian. 1971. Perception and awareness of air pollution in Toronto. Natural Hazard Research Working Paper no. 13. Toronto: University of Toronto, Department of Geography.

Backstrom, Charles H., and Hursh, Gerald D. 1963. Survey research. Evanston, Ill.: Northwestern University Press.

Ballweg, John A. 1972. Measuring attitudes toward water use priorities. Water Resources Research Center Bulletin no. 50. Blackburg, Va.: Virginia Polytechnic Institute.

Barker, Mary L. 1974. Information and complexity: The conceptualisation of air pollution by specialist groups. Environment and Behavior 6, 3 (September): 346-77.

Baron, Norman; James, Cecil E.; Tideman, Philip; and Ludwig, James, eds. 1972. A survey of attitudes towards the Mississippi River as a total resource in Minnesota. Water Resources Research Center Bulletin no. 55. Minneapolis: University of Minnesota.

Bartholomew, Robert. 1974. The sonic environment and human behavior. Exchange Bibliography no. 565. Monticello, Ill.: Council of Planning Librarians.

Bauer, Raymond A. 1966. Social indicators. Cambridge: M.I.T. Press.

Baumann, Duane E., and Kasperson, Roger E. 1974. Public acceptance of renovated waste water: Myth and reality. Water Resources Research 10, 4 (August): 667-74.

Berry, Brian J. L., ed. 1977. The social burdens of environmental pollution. Cambridge, Mass.: Ballinger.

Bishop, Dwight E. 1966. Judgements of the relative and absolute acceptability of aircraft noise. Journal of the Accoustical Society of America 40, 1 (July): 108-22.

Blalock, Hubert M., Jr. 1972. Social statistics. New York: McGraw-Hill.

Bolt, Beranek and Newman, Inc. 1967. Noise in urban and suburban areas: Results of field studies. Sponsored by U.S. Department of Housing and Urban Development. Washington, D.C.: Government Printing Office.

_____. 1970. Chicago urban noise study. Report no. 1411. Downers Grove, Ill.: Bolt, Beranek and Newman.

_____. 1971. Noise from construction equipment and operations, building equipment, and home appliances. Sponsored by U.S. Environmental Protection Agency Office of Noise Abatement and Control. Washington, D.C.: Government Printing Office.

Borsky, P. N. 1970. The use of social surveys for measuring community response to noise environments. In Transportation noises, edited by J. D. Chalupnik, pp. 219-27.

Bragdon, Clifford R. 1970. Noise pollution: The unquiet crisis. Philadelphia: University of Pennsylvania Press.

Brehm, J., and Cohen, A. 1962. Explorations in cognitive dissonance. New York: Wiley.

Bridgeland, William M., and Sofranko, Andrew J. 1975. Community structure and issue-specific influences: Community mobilization over environmental quality. Urban Affairs Quarterly 11, 2 (December): 186-214.

Brown R. 1962. Models of attitude change. In New directions in psychology, edited by R. Brown, E. Galanter, E. Hess, and G. Mandler, pp. 1-85. New York: Holt, Rinehart and Winston.

Bruvold, William H. 1971. Affective response toward uses of reclaimed water. Journal of Applied Psychology 55, 1 (February): 28-33.

_____. 1973. Belief and behavior as determinants of environmental attitudes. Environment and Behavior 5, 2 (June): 202-18.

Buchanan, J. M. 1968. A behavioral theory of pollution. Western Economic Journal 6: 347-58.

Burns, William. 1968. Noise and man. London: John Murray.

Cameron, Paul; Robertson, Donald; and Zaks, Jeffry. 1972. Sound pollution, noise pollution and health: Community parameters. Journal of Applied Psychology 56, 1 (February): 67-74.

Carp, Frances M.; Zawadski, Rick T.; and Shokrkon, Hossein. 1976. Dimensions of urban environmental quality. Environment and Behavior 8, 2 (June): 239-63.

Cederloff, R.; Johnsson, E.; and Kajland, A. 1963. Annoyance reactions to noise from motor vehicles: An experimental study. *Acustica* 13: 270-79.

Clark, W. E., and Pietrasanta, A. C. 1961. Community and industrial noise. *American Journal of Public Health* 51: 1329-37.

Committee on the problem of noise. 1963. *Noise: Final report presented to Parliament*. Vol. 22. London: Her Majesty's Stationery Office.

Cooke, Ronald U., and Saarinen, Thomas F. 1971. Public perception of environmental quality in Tucson, Arizona. *Journal of the Arizona Academy of Science* 6, 4 (June): 260-74.

Costantini, Edmond, and Hanf, Kenneth. 1972. Environmental concern and Lake Tahoe: A study of elite perceptions, backgrounds, and attitudes. *Environment and Behavior* 4, 2 (June): 209-42.

Coughlin, Robert E. 1975. *The perception and valuation of water quality: A review of research method and findings*. Regional Science Research Institute Discussion Paper Series no. 80. Philadelphia: University of Pennsylvania.

Craik, Kenneth H. 1973. Environmental psychology. *Annual Review of Psychology* 24:403-22.

Creer, R. N.; Gray, R. M.; and Treshow, M. 1970. Differential responses to air pollution as an environmental health problem. *Journal of the Air Pollution Control Association* 20 (12): 814-18.

Crowe, M. Jay. 1968. Toward a "definitional model" of public perceptions of air pollution. *Journal of the Air Pollution Control Association* 18 (3): 154-57.

Dasgupta, Satadal. 1967. *Attitudes of local residents toward watershed development*. Social Science Research Center Preliminary Report no. 18. State College: Mississippi State University, Water Resources Research Institute.

David, Elizabeth L. 1971. Public perception of water quality. *Water Resources Research* 7, (3): 453-57.

——————. 1974. The role of the public in decision making. In *Priorities in water management*, edited by F. M. Leversedge. Western Geographical Series, vol. 8. Victoria, B.C.: University of Victoria Press.

De Groot, Ido. 1967. Trends in public attitudes toward air pollution. *Journal of the Air Pollution Control Association* 17 (10): 679-81.

De Groot, Ido; Loring, William; Rihm, Alexander; Samuels, Sheldon W.; and Winkelstein, Warren. 1966. People and air pollution: A study of attitudes in Buffalo, New York. *Journal of the Air Pollution Control Association* 16 (5): 245-47.

Dillehay, Ronald C.; Brufold, William H.; and Siegel, Jacob P. 1969. Attitude, object label and stimulus factors in response to an attitude object. *Journal of Personality and Social Psychology* 11 (March): 220-23.

Ditton, Robert B., and Goodale, Thomas L. 1974. Water quality perceptions and attitudes. *Journal of Environmental Education* 6, 2 (winter): 21-27.

Doob, L. W. 1947. The behavior of attitudes. *Psychological Review* 54:135-56.

Dulany, D. E. 1961. Hypotheses and habits in verbal operant conditioning. *Journal of Abnormal and Social Psychology* 63:251-63.

_____. 1968. Awareness, rules, and propositional control: A confrontation with S-R behavior theory. In *Verbal behavior and S-R behavior theory*, edited by D. Horton and T. Dixon, pp. 340-87. Englewood Cliffs, N.J.: Prentice-Hall.

Dunlap, Riley E. 1975. *Sociological and social-psychological perspectives on environmental issues: A bibliography*. Exchange bibliography no. 916. Monticello, Ill.: Council of Planning Librarians.

Dynes, Russel R., and Wenger, Dennis. 1971. Factors in the community perception of water resources problems. *Water Resources Bulletin* 7 (August): 644-51.

Edwards, W. 1954. The theory of decision making. *Psychological Bulletin* 51:380-417.

Festinger, L. 1957. *A theory of cognitive dissonance*. Stanford: Stanford University Press.

Finnie, William C. 1973. Field experiments in litter control. *Environment and Behavior* 5, 2 (June): 123-44.

Fishbein, Martin. 1967a. A consideration of beliefs and their role in attitude measurement. In *Readings in attitude theory and measurement*, edited by M. Fishbein, pp. 257-66. New York: Wiley.

_____. 1967b. *Readings in attitude theory and measurement*. New York: Wiley.

_____. 1973. The prediction of behavior from attitudinal variables. In *Advances in communication research*, edited by C. D. Mortensen and K. Sereno, pp. 3-31. New York: Harper and Row.

Fishbein, Martin, and Ajzen, Icek. 1975. *Belief, attitude, intention and behavior: An introduction to theory and research*. Reading, Pa.: Addison-Wesley.

Foreman, J. E. K.; Emmerson, M. A.; and Dickinson, S. M. 1974. Noise level attitudinal surveys of London and Woodstock, Ontario. *Sound and Vibration* 8, 12 (December): 16-22.

Frederickson, H. George, and Magnas, Howard. 1968. Comparing attitudes toward water pollution in Syracuse. *Water Resources Research* 4 (5): 877-89.

Geller, E. Scott; Witmer, Jill F.; and Orebaugh, Andra L. 1976. Instructions as a determinant of paper-disposal behaviors. Environment and Behavior 8, 3 (September): 417-39.

Glass, David C., and Singer, Jerome E. 1972. Urban stress: Experiments on noise and social stressors. New York: Academic Press.

Goodchild, Barry. 1974. Class differences in environmental perception: An exploratory study. Urban Studies 11, 2 (June): 157-69.

Goodfriend, L. S., Associates. 1971. Noise from industrial plants. Sponsored by the U.S. Environmental Protection Agency, Office of Noise Abatement and Control. Washington, D.C.: Government Printing Office.

Goodman, Robert F., and Clary, Bruce B. 1976. Community attitudes and action in response to airport noise. Environment and Behavior 8, 3 (September): 441-70.

Gore, Peter H.; Wilson, Samuel; and Capener, Harold R. 1975. A sociological approach to the problem of water pollution. Growth and Change 6 (January): 17-22.

Griffiths, I. D., and Langdon, F. J. 1968. Subjective response to road traffic noise. Journal of Sound and Vibration 8, (1): 16-32.

Heider, F. 1944. Social perception and phenomenal causality. Psychological Review 51:358-74.

_____. 1946. Attitudes and cognitive organization. Journal of Psychology 21:107-12.

_____. 1958. The Psychology of interpersonal relations. New York: Wiley.

Hovland, C. I.; Janis, I. L.; and Kelley, H. H. 1953. Communication and persuasion. New Haven: Yale University Press.

Hull, C. L. 1943. The principles of behavior. New York: Appleton-Century-Crofts.

_____. 1951. Essentials of behavior. New Haven: Yale University Press.

Hunter, Albert. 1974. Symbolic communities: The persistence and change of Chicago's local communities. Chicago: University of Chicago Press.

Ibsen, Charles A., and Ballweg, John A. 1969. Public perceptions of water resource problems. Water Resources Research Center Bulletin no. 29. Blackburg, Va.: Virginia Polytechnic Institute.

Illinois report probes citizens' attitudes on refuse problems. U.S. Environmental Protection Agency reprint. Washington, D.C.: Government Printing Office.

Insko, Chester A. 1967. Theories of attitude change. Englewood Cliffs, N.J.: Prentice-Hall.

Jacoby, Louis R. 1972. Perception of air, noise and water Pollution in Detroit. Michigan Geographical Publication no. 7. Ann Arbor: University of Michigan Department of Geography.

Johnson, Dale L. 1972. Air pollution: Public attitudes and public action. American Behavioral Scientist 15, 4 (March-April): 533-61.

Johnson, K. L. 1968. Citizen complaints of air pollution in northeastern Illinois. Journal of the Air Pollution Control Association 18, 6: 399-401.

Johnson, R. J., and Hay, J. E. 1974. Spatial variations in awareness of air pollution distributions. International Journal of Environmental Studies 6 (May-July): 131-36.

Jonsson, E. 1963. Annoyance reactions to external environmental factors in different sociological groups. Acta Sociologica 7: 229-59.

Jonsson, E.; Kauland, A; Paccagnella, B.; and Sörensen, S. 1969. Annoyance reactions to traffic noise in Italy and Sweden. Archives of Environmental Health 19, 11 (November): 692-99.

Jonsson, E., and Sörensen, S. 1975. Reliability of forecasts of annoyance reactions. Archives of Environmental Health 30, 2 (February): 104-6.

Katz, Daniel. 1960. The functional approach to the study of attitudes. Public Opinion Quarterly 24:163-204.

Klee, A. J. 1971. Aspects of sampling attitudes towards solid waste programs. Sponsored by Office of Solid Waste Management Programs, U.S. Environmental Protection Agency. Washington, D.C.: Government Printing Office.

Koczkur, Eugene; Broger, Eric; Henderson, Valtin; and Lightstone, Alfred. 1973. Noise monitoring and a sociological survey in the city of Toronto. Journal of the Air Pollution Control Association 23, 2 (February): 105-9.

Kohan, Stephanie; DeMille, Richard; and Myers, James H. 1972. Two comparisons of attitude measures. Journal of Advertising Research 12, 4 (August): 29-34.

Kromm, David E. 1973. Response to air pollution in Ljubljana Yugoslavia. Annals of the American Association of Geographers 63 (June): 208-17.

Kromm, David E.; Probald, F.; and Wall, G. 1973. An international comparison of response to air pollution. Journal of Environmental Management 1:363-75.

Kryter, Karl D. 1970. The effects of noise on man. New York: Academic Press.

Langowski, Alan, and Sigler, Jeanne. 1971. Citizen attitudes toward the environment: An appraisal of the research. Illinois Institute for Environmental Quality Final Report. Chicago: Illinois Institute for Environmental Quality.

Liu, Ben-chieh. 1975. Quality of life indicators in U.S. Metropolitan areas, 1970. Sponsored by the U.S. Environmental Protection Agency. Washington, D.C.: Government Printing Office.

Loether, Herman J., and McTavish, Donald G. 1974a. Descriptive statistics for sociologists. Boston: Allyn and Bacon.

_____. 1974b. Inferential statistics for sociologists. Boston: Allyn and Bacon.

Lycan, D. R., and Sewell, W. R. D. 1968. Water and air pollution as components of the urban environment of Victoria. In Geographical perspectives: Some northern viewpoints, edited by G. S. Tomkins, pp. 13-18. British Columbia Geographical Series no. 8. Vancouver: Tantalus Press.

McKennell, A. C. 1970. Noise complaints and community action. In Transportation noises, edited by J. D. Chalupnik, pp. 228-33. Seattle: University of Washington Press.

McMeiken, J. Elizabeth. 1972. Public health professionals and the environment: A study of perceptions and attitudes. Department of Environment, Inland Waters Directorate, Water Planning and Management Branch, Social Science Series no. 5. Ottawa: John Lovell.

McMeiken, J. Elizabeth, and Rostron, J. 1970. Spatial variations in the perception of pollution: A pilot study on Vancouver Island, B.C. In The geographer and society, edited by Derrick Sewell and Harold D. Foster, pp. 79-95. Western Geographical Series no. 1. Victoria: University of Victoria Department of Geography.

McMullen, Thomas B.; Fensterstock, Jack C.; Faoro, Robert B.; and Smith, Raymond. 1968. Air quality and characteristic community parameters. Journal of the Air Pollution Control Association 18, 8 (August): 545-49.

Medalia, N. Z. 1964. Air pollution as a socio-environmental health problem: A survey report. Journal of Health and Human Behavior 5, 4 (Winter): 154-65.

Medalia, N. Z., and Finkner, A. L. 1965. Community perception of air quality: An opinion survey in Clarkston, Washington. Sponsored by U.S. Public Health Service. Washington, D.C.: Government Printing Office.

Mehrabian, Albert, and Russell, James A. 1974. An approach to environmental psychology. Cambridge: MIT Press.

Meyer, Philip. 1973. Precision journalism. Bloomington: Indiana University Press.

Milbrath, L., and Sahr, R. 1975. Perceptions of environmental quality. Social Indicators Research 1:397-438.

Mitchell, Bruce. 1971. Behavioral aspects of water management: A paradigm and a case study. Environment and Behavior 3, 2 (June): 135-53.

Moos, Rudolf H., and Insel, Paul M. 1974. Issues in social ecology: Human milieus. Palo Alto, Calif.: National Press Books.

National Analysts, Inc. 1973. Metropolitan housewives' attitudes toward solid waste disposal. U.S. Environmental Protection Agency Report no. EPA-R5-72-003. Washington, D.C.: Government Printing Office.

Newcomb, T. M. 1953. An approach to the study of communicative acts. Psychological Review 60:393-404.

_____. 1959. Individual systems of orientation. In Psychology: A study of science, edited by S. Koch, 3:384-422. New York: McGraw-Hill.

Nixon, Charles W., and Borsky, Paul N. 1966. Effects of sonic boom on people: St. Louis, Missouri, 1961-1962. Journal of the Acoustical Society of America 39, 5 (May): 951-58.

O'Riordan, Timothy. 1971. Public opinion and environmental quality. Environment and Behavior 3, 2 (June): 191-214.

_____. 1973. Some reflections on environmental attitudes and behavior. Area 5 (1): 17-21.

Osgood, C. E., and Tannenbaum, P. H. 1955. The principle of congruity in the prediction of attitude change. Psychological Review 62:42-55.

Pagorski, A. D. 1974. Is the public ready for recycled water? Water and Sewage Works 121, 6 (June): 108-9.

Parkes, J. G. M. 1974. User response to water quality and water based recreation in the Qu'Appelle Valley, Saskatchewan. In Priorities in water management, edited by F. M. Leversedge, pp. 99-112. Western Goegraphical Series, vol. 8. Victoria: University of Victoria Press.

Proshansky, H. M.; Ittleson, W. H.; and Rivlin, L. G. 1970. Environmental psychology: Man and his physical setting. New York: Holt, Rinehart and Winston.

Rankin, Robert E. 1969. Air pollution control and public policy. Journal of the Air Pollution Control Association 19 (8): 565-69.

Rao, Potluri, and Miller, Roger L. 1971. Applied econometrics. Belmont, Calif.: Wadsworth.

Reid, Dennis H.; Luyben, Paul D.; Rawers, Robert J.; and Bailey, Jon S. 1976. Newspaper recycling behavior: The effects of prompting and proximity of containers. Environment and Behavior 8, 3 (September): 471-82.

Rich, Thomas. 1974. Pollution perception and political response case study: Ebensburg, Pennsylvania. Pennsylvania Geographer 12, 2 (July): 23-28.

Robinson, D. W.; Bowsher, J. M.; and Copeland, W. C. 1963. On judging the noise from aircraft in flight. In Noise: Final report, pp. 186-203. London: Her Majesty's Stationery Office.

Robinson, John P., and Shaver, Phillip R. 1973. Measures of social psychological attitudes. Ann Arbor: University of Michigan Institute for Social Research.

Robinson, Stuart N. 1976. Littering behavior in public places. *Environment and Behavior* 8, 3 (September): 363-84.

Rokeach, Milton. 1968. *Beliefs, attitudes, and values: A theory of organization and change*. San Francisco: Jossey-Bass.

Rosenberg, Milton. 1956. Cognitive structure and attitudinal affect. *Journal of Abnormal and Social Psychology* 53:367-72.

_____. 1960. An analysis of affective-cognitive consistency. In *Attitude organization and change*, edited by C. Hovland and M. Rosenberg, pp. 15-64. New Haven: Yale University Press.

Rosenberg, Milton, and Abelson, R. 1960. An analysis of cognitive balancing. In *Attitude organization and change*, edited by C. Hovland and M. Rosenberg, pp. 112-63. New Haven: Yale University Press.

Saarinen, Thomas F. 1969. *Perception of environment*. Commission on College Geography Resource Paper no. 5. Washington, D.C.: Association of American Geographers.

Schusky, Jane. 1966. Public awareness and concern with air pollution in the St. Louis metropolitan area. *Journal of the Air Pollution Control Association* 16 (2): 72-76.

Schusky, Jane; Goldner, Lester; Mann, Seymour; and Loring, William. 1964. Methodology for the study of public attitudes concerning air pollution. *Journal of the Air Pollution Control Association* 14 (11): 445-58.

Sewell, W. R. Derrick. 1974. Water resources planning and policy making: Challenges and responses. In *Priorities in water management*, edited by F. M. Leversedge, pp. 259-86. Western Geographical Series, vol. 8. Victoria, B.C.: University of Victoria Press.

Sewell, W. R. Derrick, and Burton, Ian. 1971. *Perceptions and attitudes in resources management*. Policy Research and Coordination Branch, Resource Paper no. 2. Ottawa, Canada: Department of Energy, Mines and Resources.

Sigler, Jeanne. 1973. *Attitudes of Illinois citizens toward solid waste and the environment*. Illinois Institute for Environmental Quality, Final Report no. 73-6. Chicago: Illinois Institute for Environmental Quality.

Sims, John, and Baumann, Duane. 1974a. *Human behavior and the environment: Interactions between man and his physical world*. Chicago: Maaroufa Press.

_____. 1974b. Renovated waste water: The question of public acceptance. *Water Resources Research* 10 (4): 659-66.

_____. 1976. Professional bias and water reuse. *Economic Geography* 52, 1 (January): 1-10.

Smith, W. S.; Zeidberg, L. D.; and Schuenemann, J. J. 1964. Public reaction to air pollution in Nashville, Tennessee. *Journal of the Air Pollution Control Association* 14 (10): 418-23.

Sorenson, S., et al. 1974. Interview and mailed questionnaires for the evaluation of annoyance reactions. Environmental Research 8 (October): 166-70.

Southern Illinois University. 1965. Public awareness and concern with air pollution in the St. Louis metropolitan area. U.S. Public Health Service Report PH86-63-131. Washington, D.C.: Government Printing Office.

Staats, A. W. 1968. Social behaviorism and human motivation: Principles of the attitude-reinforcer-discriminative system. In Psychological foundations of attitudes, edited by A. G. Greenwald, T. C. Brock, and T. M. Ostrom, pp. 33-66. New York: Academic Press.

Stalker, W. W., and Robinson, C. B. 1967. A method for using air pollution measurements and public opinion to establish ambient air quality standards. Journal of the Air Pollution Control Association 17 (4): 142-44.

Swan, James A. 1970. Response to air pollution: A study of attitudes and coping strategies of high school youths. Environment and Behavior 2, 2 (September): 127-51.

_____. 1972. Public response to air pollution. In Environment and the social sciences: Perspectives and applications, edited by J. F. Wohlwill and D. H. Carson, pp. 66-74. New York: American Psychological Association.

Thomas, W. I. 1966. Social personality: Organization of attitudes. In Social organization and social personality, edited by M. Janowitz, pp. 11-36. Chicago: University of Chicago Press.

Thomas, William A., ed. 1972. Indicators of environmental quality. New York: Plenum.

Thurstone, L. L. 1931. The measurement of attitudes. Journal of Abnormal and Social Psychology 26:249-69.

Tognacci, Louis N.; Weigel, Russell H.; Wideen, Marvin F.; and Vernon, David T. A. 1972. Environmental quality: How universal is public concern? Environment and Behavior 4, 1 (March): 73-86.

Tolman, E. C. 1932. Purposive behavior in animals and men. New York: Appleton-Century-Crofts.

Triandis, H. C. 1971. Attitudes and attitude change. New York: Wiley.

Unwin, D. J., and Holtby, F. E. 1974. Public perception of cleaner air in the Manchester area, 1970. Cambria 1 (1): 43-51.

Van Arsdol, Maurice; Sabagh, Georges; and Alexander, Francesca. 1964. Reality and the perception of environmental hazards. Journal of Health and Human Behavior 5:144-53.

Vlachos, E. 1975. Social aspects of solid wastes development and management: Refuse, recovery, reuse. Water, Air, and Soil Pollution 4, 2 (May): 293-301.

Wall, Geoffrey. 1973a. Public response to air pollution in South Yorkshire, England. Environment and Behavior 5, 2 (June): 219-48.

_____. 1973b. Public response to air pollution in Sheffield, England. International Journal of Environmental Studies 5, 4 (December): 259-70.

Watkins, George A. 1974. Developing a "water concern" scale. Journal of Environmental Education 5, 4 (summer): 54-58.

Weinstein, Neil David. 1976. The statistical prediction of environmental preferences: Problems of validity and application. Environment and Behavior 8, 4 (December): 611-26.

White, Gilbert F. 1966. Formation and role of public attitudes. In Environmental quality in a growing economy, edited by Henry Jarrett, pp. 105-27. Baltimore: Johns Hopkins University Press, for Resources for the Future.

_____. 1973a. Natural hazard perception and choice. London: Oxford University Press.

_____. 1973b. Natural hazards research. In Directions in geography, edited by Richard J. Chorley, pp. 193-216. London: Methuen.

Williams, J. D., and Bunyard, F. L. 1966. Interstate air pollution study: Phase II project report--opinion surveys and air quality statistical relationships.

Winham, G. 1972. Attitudes on pollution and growth in Hamilton; or, There's an awful lot of talk these days about ecology. Canadian Journal of Political Science 5 (September): 389-401.

Zajonc, R. 1960. The concepts of balance, congruity and dissonance. Public Opinion Quarterly 24:280-96.

THE UNIVERSITY OF CHICAGO
DEPARTMENT OF GEOGRAPHY
RESEARCH PAPERS (Lithographed, 6×9 Inches)

(Available from Department of Geography, The University of Chicago, 5828 S. University Ave., Chicago, Illinois 60637. Price: $6.00 each; by series subscription, $5.00 each.)

106. SAARINEN, THOMAS F. *Perception of the Drought Hazard on the Great Plains* 1966. 183 pp.
107. SOLZMAN, DAVID M. *Waterway Industrial Sites: A Chicago Case Study* 1967. 138 pp.
108. KASPERSON, ROGER E. *The Dodecanese: Diversity and Unity in Island Politics* 1967. 184 pp.
109. LOWENTHAL, DAVID, et al. *Environmental Perception and Behavior.* 1967. 88 pp.
110. REED, WALLACE E. *Areal Interaction in India: Commodity Flows of the Bengal-Bihar Industrial Area* 1967. 210 pp.
112. BOURNE, LARRY S. *Private Redevelopment of the Central City: Spatial Processes of Structural Change in the City of Toronto* 1967. 199 pp.
113. BRUSH, JOHN E., and GAUTHIER, HOWARD L., JR. *Service Centers and Consumer Trips: Studies on the Philadelphia Metropolitan Fringe* 1968. 182 pp.
114. CLARKSON, JAMES D. *The Cultural Ecology of a Chinese Village: Cameron Highlands, Malaysia* 1968. 174 pp.
115. BURTON, IAN; KATES, ROBERT W.; and SNEAD, RODMAN E. *The Human Ecology of Coastal Flood Hazard in Megalopolis* 1968. 196 pp.
117. WONG, SHUE TUCK. *Perception of Choice and Factors Affecting Industrial Water Supply Decisions in Northeastern Illinois* 1968. 96 pp.
118. JOHNSON, DOUGLAS L. *The Nature of Nomadism* 1969. 200 pp.
119. DIENES, LESLIE. *Locational Factors and Locational Developments in the Soviet Chemical Industry* 1969. 285 pp.
120. MIHELIC, DUSAN. *The Political Element in the Port Geography of Trieste* 1969. 104 pp.
121. BAUMANN, DUANE. *The Recreational Use of Domestic Water Supply Reservoirs: Perception and Choice* 1969. 125 pp.
122. LIND, AULIS O. *Coastal Landforms of Cat Island, Bahamas: A Study of Holocene Accretionary Topography and Sea-Level Change* 1969. 156 pp.
123. WHITNEY, JOSEPH. *China: Area, Administration and Nation Building* 1970. 198 pp.
124. EARICKSON, ROBERT. *The Spatial Behavior of Hospital Patients: A Behavioral Approach to Spatial Interaction in Metropolitan Chicago* 1970. 198 pp.
125. DAY, JOHN C. *Managing the Lower Rio Grande: An Experience in International River Development* 1970. 277 pp.
126. MAC IVER, IAN. *Urban Water Supply Alternatives: Perception and Choice in the Grand Basin, Ontario* 1970. 178 pp.
127. GOHEEN, PETER G. *Victorian Toronto, 1850 to 1900: Pattern and Process of Growth* 1970. 278 pp.
128. GOOD, CHARLES M. *Rural Markets and Trade in East Africa* 1970. 252 pp.
129. MEYER, DAVID R. *Spatial Variation of Black Urban Households* 1970. 127 pp.
130. GLADFELTER, BRUCE. *Meseta and Campiña Landforms in Central Spain: A Geomorphology of the Alto Henares Basin* 1971. 204 pp.
131. NEILS, ELAINE M. *Reservation to City: Indian Urbanization and Federal Relocation* 1971. 200 pp.
132. MOLINE, NORMAN T. *Mobility and the Small Town, 1900–1930* 1971. 169 pp.
133. SCHWIND, PAUL J. *Migration and Regional Development in the United States, 1950–1960* 1971. 170 pp.
134. PYLE, GERALD F. *Heart Disease, Cancer and Stroke in Chicago: A Geographical Analysis with Facilities Plans for 1980* 1971. 292 pp.
135. JOHNSON, JAMES F. *Renovated Waste Water: An Alternative Source of Municipal Water Supply in the U.S.* 1971. 155 pp.
136. BUTZER, KARL W. *Recent History of an Ethiopian Delta: The Omo River and the Level of Lake Rudolf* 1971. 184 pp.
137. HARRIS, CHAUNCY D. *Annotated World List of Selected Current Geographical Serials in English, French, and German* 3rd edition 1971. 77 pp.
138. HARRIS, CHAUNCY D., and FELLMANN, JEROME D. *International List of Geographical Serials* 2nd edition 1971. 267 pp.
139. MC MANIS, DOUGLAS R. *European Impressions of the New England Coast, 1497–1620* 1972. 147 pp.
140. COHEN, YEHOSHUA S. *Diffusion of an Innovation in an Urban System: The Spread of Planned Regional Shopping Centers in the United States, 1949–1968* 1972. 136 pp.

141. MITCHELL, NORA. *The Indian Hill-Station: Kodaikanal* 1972. 199 pp.
142. PLATT, RUTHERFORD H. *The Open Space Decision Process: Spatial Allocation of Costs and Benefits* 1972. 189 pp.
143. GOLANT, STEPHEN M. *The Residential Location and Spatial Behavior of the Elderly: A Canadian Example* 1972. 226 pp.
144. PANNELL, CLIFTON W. *T'ai-chung, T'ai-wan: Structure and Function* 1973. 200 pp.
145. LANKFORD, PHILIP M. *Regional Incomes in the United States, 1929–1967: Level, Distribution, Stability, and Growth* 1972. 137 pp.
146. FREEMAN, DONALD B. *International Trade, Migration, and Capital Flows: A Quantitative Analysis of Spatial Economic Interaction* 1973. 202 pp.
147. MYERS, SARAH K. *Language Shift Among Migrants to Lima, Peru* 1973. 204 pp.
148. JOHNSON, DOUGLAS L. *Jabal al-Akhdar, Cyrenaica: An Historical Geography of Settlement and Livelihood* 1973. 240 pp.
149. YEUNG, YUE-MAN. *National Development Policy and Urban Transformation in Singapore: A Study of Public Housing and the Marketing System* 1973. 204 pp.
150. HALL, FRED L. *Location Criteria for High Schools: Student Transportation and Racial Integration* 1973. 156 pp.
151. ROSENBERG, TERRY J. *Residence, Employment, and Mobility of Puerto Ricans in New York City* 1974. 230 pp.
152. MIKESELL, MARVIN W., editor. *Geographers Abroad: Essays on the Problems and Prospects of Research in Foreign Areas* 1973. 296 pp.
153. OSBORN, JAMES. *Area, Development Policy, and the Middle City in Malaysia* 1974. 273 pp
154. WACHT, WALTER F. *The Domestic Air Transportation Network of the United States* 1974. 98 pp.
155. BERRY, BRIAN J. L., et al. *Land Use, Urban Form and Environmental Quality* 1974. 464 pp.
156. MITCHELL, JAMES K. *Community Response to Coastal Erosion: Individual and Collective Adjustments to Hazard on the Atlantic Shore* 1974. 209 pp.
157. COOK, GILLIAN P. *Spatial Dynamics of Business Growth in the Witwatersrand* 1975. 143 pp.
158. STARR, JOHN T., JR. *The Evolution of Unit Train Operations in the United States: 1960–1969—A Decade of Experience* 1976. 247 pp.
159. PYLE, GERALD F. *The Spatial Dynamics of Crime* 1974. 220 pp.
160. MEYER, JUDITH W. *Diffusion of an American Montessori Education* 1975. 109 pp.
161. SCHMID, JAMES A. *Urban Vegetation: A Review and Chicago Case Study* 1975. 280 pp.
162. LAMB, RICHARD. *Metropolitan Impacts on Rural America* 1975. 210 pp.
163. FEDOR, THOMAS. *Patterns of Urban Growth in the Russian Empire during the Nineteenth Century* 1975. 275 pp.
164. HARRIS, CHAUNCY D. *Guide to Geographical Bibliographies and Reference Works in Russian or on the Soviet Union* 1975. 496 pp.
165. JONES, DONALD W. *Migration and Urban Unemployment in Dualistic Economic Development* 1975. 186 pp.
166. BEDNARZ, ROBERT S. *The Effect of Air Pollution on Property Value in Chicago* 1975. 118 pp.
167. HANNEMANN, MANFRED. *The Diffusion of the Reformation in Southwestern Germany, 1518–1534* 1975. 248 pp.
168. SUBLETT, MICHAEL D. *Farmers on the Road. Interfarm Migration and the Farming of Noncontiguous Lands in Three Midwestern Townships, 1939–1969* 1975. 228 pp.
169. STETZER, DONALD FOSTER. *Special Districts in Cook County: Toward a Geography of Local Government* 1975. 189 pp.
170. EARLE, CARVILLE V. *The Evolution of a Tidewater Settlement System: All Hallow's Parish, Maryland, 1650–1783* 1975. 249 pp.
171. SPODEK, HOWARD. *Urban-Rural Integration in Regional Development: A Case Study of Saurashtra, India—1800–1960* 1976. 156 pp.
172. COHEN, YEHOSHUA S. and BERRY, BRIAN J. L. *Spatial Components of Manufacturing Change* 1975. 272 pp.
173. HAYES, CHARLES R. *The Dispersed City: The Case of Piedmont, North Carolina* 1976. 169 pp.
174. CARGO, DOUGLAS B. *Solid Wastes: Factors Influencing Generation Rates* 1977. 112 pp.
175. GILLARD, QUENTIN. *Incomes and Accessibility. Metropolitan Labor Force Participation, Commuting, and Income Differentials in the United States, 1960–1970* 1977. 140 pp.
176. MORGAN, DAVID J. *Patterns of Population Distribution: A Residential Preference Model and Its Dynamic* 1978. 216 pp.
177. STOKES, HOUSTON H.; JONES, DONALD W. and NEUBURGER, HUGH M. *Unemployment and Adjustment in the Labor Market: A Comparison between the Regional and National Responses* 1975. 135 pp.

178. PICCAGLI, GIORGIO ANTONIO. *Racial Transition in Chicago Public Schools. An Examination of the Tipping Point Hypothesis, 1963-1971* 1977.
179. HARRIS, CHAUNCY D. *Bibliography of Geography. Part I. Introduction to General Aids* 1976. 288 pp.
180. CARR, CLAUDIA J. *Pastoralism in Crisis. The Dasanetch and their Ethiopian Lands.* 1977. 339 pp.
181. GOODWIN, GARY C. *Cherokees in Transition: A Study of Changing Culture and Environment Prior to 1775.* 1977. 221 pp.
182. KNIGHT, DAVID B. *A Capital for Canada: Conflict and Compromise in the Nineteenth Century.* 1977. 359 pp.
183. HAIGH, MARTIN J. *The Evolution of Slopes on Artificial Landforms: Blaenavon, Gwent.* 1978. 311 pp.
184. FINK, L. DEE. *Listening to the Learner. An Exploratory Study of Personal Meaning in College Geography Courses.* 1977. 200 pp.
185. HELGREN, DAVID M. *Rivers of Diamonds: An Alluvial History of the Lower Vaal Basin.* 1978.
186. BUTZER, KARL W., editor. *Dimensions of Human Geography: Essays on Some Familiar and Neglected Themes.* 1978. 201 pp.
187. MITSUHASHI, SETSUKO. *Japanese Commodity Flows.* 1978. 185 pp.
188. CARIS, SUSAN L. *Community Attitudes toward Pollution.* 1978. 226 pp.